# THE DIRTY DOZEN

# THE DIRTY DOZEN

## Toxic Chemicals and the Earth's Future

*Bruce E. Johansen*

Westport, Connecticut
London

**Library of Congress Cataloging-in-Publication Data**

Johansen, Bruce E. (Bruce Elliott), 1950–
    The dirty dozen : toxic chemicals and the earth's future / Bruce E.
Johansen
        p. cm.
    Includes bibliographical references and index.
    ISBN 0–275–97702–1 (alk. paper)
        1. Organochlorine compounds—Environmental aspects.
    2. Persistent pollutants—Environmental aspects.   3. Organochlorine
    compounds—Health aspects.   4. Persistent pollutants—Health
    aspects.   I. Title.
TD196.C5J65   2003
363.738′4—dc21        2002029872

British Library Cataloguing in Publication Data is available.

Library of Congress Catalog Card Number: 2002029872
ISBN: 0–275–97702–1

First published in 2003

Praeger Publishers, 88 Post Road West, Westport, CT 06881
An imprint of Greenwood Publishing Group, Inc.
www.praeger.com

Printed in the United States of America

(∞)™

The paper used in this book complies with the
Permanent Paper Standard issued by the National
Information Standards Organization (Z39.48–1984).

10 9 8 7 6 5 4 3

**Copyright Acknowledgments**

The author and publisher gratefully acknowledge permission for use of
the following material:

Excerpts from the personal communications of Sheila Watt-Cloutier.

Excerpts from the personal communications of Barbara Alice Mann.

# Contents

# Preface

I was introduced to the human and natural toll of persistent organic pollutants in a scholarly way via e-mailed descriptions of a press conference held in New York City with, among others, the eminent ecologist Barry Commoner, and Sheila Watt-Cloutier, Canadian president of the Inuit Circumpolar Conference, October 3, 2000 (Commoner 1966, 1971, 1990).

Watt-Cloutier herself had become a major driving force behind the negotiation of the Stockholm Convention that will eventually ban the most widely used and dangerous organochlorines. Her testimony about the toll of persistent organic pollutants on the Inuit and Arctic animals haunted me as I plunged deeper into the subject during ensuing months. Ten months later, we met on Baffin Island, at Iqaluit, to discuss the toll of persistent organic pollutants, as well as global warming, on her people.

Commoner's study included precise identification of dioxin sources across North America that were poisoning the Inuit, and Watt-Cloutier warned that the toxicity that befalls the Inuit today could afflict all humankind a few generations hence if the production of dioxins, polyvinyl chloride (PCBs), and other bioaccumulating organochlorines is not stopped. Some of the sources that the study identified were only a few miles from my home in Omaha, Nebraska, where my concern about organochlorines had been stirred informally and anecdotally for many years by the levels of agricultural pesticide and herbicide runoff in our municipal water. Levels were high enough to prompt my wife to order bottled water when we had infants in the house.

Viewed from the top of the world, where ocean and wind currents cause the dirty dozen to concentrate, the arrival of persistent organic pollutants is a direct threat to a way of life based on Arctic animal protein. Given atmospheric and biological circumstances, the Arctic—still a pristine place in many non-Inuit imaginations—has become a testing ground for the perils of environmental toxicology. As Watt-Cloutier observes in chapter 2, the lives of the Inuit have been imperiled by persistent organic pollutants in a way that many of the chemicals' producers and consumers have only slowly come to realize.

After hearing from Watt-Cloutier that many Inuit mothers were refraining from breast-feeding their children because of dioxins, PCBs (polychlorinated biphenyls), and other chemicals in their food chain, I set out to compose an account of these toxins that would be accessible to undergraduate students and other members of the reading public in the areas of North America where they are produced, before their transport by atmospheric and oceanic circulations to the Arctic. I began with the dirty dozen, the twelve substances that have been short-listed for elimination by the Stockholm Convention, including the organochlorine pesticides (DDT, chlordane, mirex, hexachlorobenzene, endrin, aldrin, dieldrin, toxaphene, heptachlor), as well as industrial chemicals (including PCBs and the supertoxic dioxins and furans). Because some of these substances (notably the PCBs and dioxins) are actually families comprising hundreds of chemicals, this list could just as aptly be called the dirty hundreds as the dirty dozen.

I read with astonishment Watt-Cloutier's descriptions of how Inuit children were being injured by exposure to PCBs and dioxins, and of Inuit mothers' toxic breast milk. Hers was an account that very few people in North America had heard at that time. Most of my undergraduate students in communication professed ignorance of the chemicals afflicting the Arctic, such as PCBs and dioxins. Only when I mentioned Agent Orange did I receive a knowing nod from students who, for the most part, dread our required courses in basic sciences, including chemistry.

Unlike the debate over global warming, about which President George W. Bush has played antagonist to most of the world, the evidence of damage (past and potential) of persistent organic pollutants is so incontrovertible that even the United States has pledged support of the Stockholm Convention. What remains to be negotiated are the methods and pace of their elimination, first in manufacturing and then, as the twenty-first century passes, in our atmosphere, oceans, and food web.

## ACKNOWLEDGMENTS

Thanks are due the staff of the *Progressive,* especially its editor-in-chief Matthew Rothschild, for publishing an article that started me down the road to this book (Johansen 2000). The *Progressive* also paid for a trip to the Arctic during the summer of 2001 that allowed me to meet Sheila Watt-Cloutier personally. I owe her my thanks for arranging interviews that allowed me to witness the perils of industry for today's Inuit—not only the ravages of dioxins and PCBs, but also the accelerating pace of global warming that is destroying the area's ice-based ecology.

I also owe a debt of gratitude, as always, to the staff of University of Nebraska at Omaha Interlibrary Loan, and to Deborah Smith-Howell, Communication Department chairwoman, for partial relief from teaching duties that has allowed me time to research this and other books and articles. This book also has benefited immensely from the comments of an anonymous peer reviewer who provided a rigorous red inking of the manuscript before it began its sojourn through production at Greenwood Press. At the press, thanks also are due my editors on this volume, Cynthia Harris and Heather Ruland Staines.

Finally, many thanks to my wife, Pat Keiffer, for her intellectual stimulation, advice, support, encouragement, and wonderful dinners following long days of slogging through some very complex, sometimes depressing, but very necessary information that will be vital to the future of the Inuit—and to the environmental health of the entire world.

## REFERENCES

Commoner, Barry. 1966. *Science and Survival.* New York: Viking Press.
———. 1971. *The Closing Circle: Nature, Man, and Technology.* New York: Knopf.
———. 1990. *Making Peace with the Planet.* New York: Pantheon Books.
Johansen, Bruce E. 2000. "Pristine No More: The Arctic, Where Mother's Milk Is Toxic." *Progressive,* December, 27–29.

# Introduction

Our economic system has been very good at developing ways to exploit chemistry for human welfare and profit. It has been less sagacious at anticipating, detecting, and dealing with chemistry's unintended consequences. In our economy (at least until recent times) novelty, convenience, and profit came first, and environmental questions often were asked later, if ever. Medical testing often occurs only after effects have become evident anecdotally.

During 1928, for example, the *New York Times* (March 15, 1929) carried on its front page a report headlined "Professor, P.Q. Card Pictures Frocks of Asbestos." Nine days later, on page 2 of the *Sunday Times*'s section 11, a feature foretold the use of asbestos in high-fashion dresses. At about the same time, other fashion-conscious consumers were buying wristwatches with radioactive radium dials that shone in the dark. Two decades later, uranium-mining companies in New Mexico and Arizona sent Navajos into the earth to mine radioactive uranium with no protective gear, as if it was coal. During their lunch breaks, the miners washed their food down with swigs of radioactive water that flowed through the mines. Lacking toilet paper, miners sometimes cleaned themselves with wads of radioactive yellowcake. Three and four decades later, a large number of former uranium miners died of lung cancer in an area where the disease had been nearly unknown.

Earthwide, human-induced pollution is unique to our time. Human beings, as a species, no longer foul only small parts of the Earth (such as specific cities or bodies of water) at a time. Tom Brokaw's "Greatest Generation" did, indeed, win World War II. The irony is that the same generation also gave us a wonder-world of synthetic chlorinated hydrocarbons that have left a devastating environmental heritage. Each of these chemicals is built on a chlorine base. "The chlorine gas produced by the chemical industry is violently reactive, combining quickly and promiscuously with any organic matter with which it comes in contact to produce a new family of compounds called organochlorines" (Thornton 2000, 203–4).

Synthetic organochlorines such as dioxins and polychlorinated biphenyls (PCBs) are a perfect vehicle for worldwide pollution because they ignore virtually all boundaries, natural and artificial. We are now chemical neighbors of the Inuit and polar bears in the Arctic, the whales in the sea, and the penguins of Antarctica. All of us now carry in our bodies trace amounts of some of the most potent toxins created by human industrial ingenuity.

These chemicals bioaccumulate (intensify in potency) along the food chain, sometimes to thousands of times their original toxicity, posing special perils to animals, including human beings, who eat meat and fish. Problems related to their toxicity are especially acute in places, such as the polar regions, where currents in the atmosphere and oceans cause organochlorines to accumulate. The bull's-eye on our industrial chemical target has become mothers of Inuit children who now pass along their bioaccumulating toxic burden to their children in breast milk. The Arctic remains pristine only to those who know nothing of how toxic synthetic poisons behave.

Today, organochlorines produced for commerce and those created unintentionally can be found in the air, lakes, oceans, soils, sediments, and animals, including humans, in every region of the planet. The effects of these substances on the environment, particularly on human health, "has emerged as one of the most compelling scientific, political and economic issues of our time" (Allsopp, Costner, and Johnston, 1995). Readers of this book may visit villages in the Arctic west of Greenland, where people have higher concentrations of PCBs than anywhere else on Earth (except victims of industrial accidents). They are at the top of a food chain composed mainly of PCB-laced polar bears, seals, and other animals.

Persistent organic pollutants (POPs), of which PCBs are one example, are not soluble in water. They dissolve readily, however, in fats and oils. Because of their resistance to degradation and affinity for fat, POPs accumulate in the body fat of living organisms and become more concentrated as they move from one creature to another onward and upward in the food chain. In this way, extremely small levels of such contaminants in water or soil can magnify into a significant hazard to predators who feed at the top of the food web such as dolphins, polar bears, herring gulls, and human beings. In Lake Ontario, for example, the tissue of herring gulls may contain 25 million times the concentration of PCBs found in the lake's water.

Imagine being a polar bear in today's world: your body is bioaccumulating enough synthetic toxins in your body fat to imperil reproduction in coming generations. You wake from a long winter's sleep hungry for seal meat, but you find that the ice shelf that once brought you food has moved offshore, a result of global warming. If you are close to a human settlement (such as Churchill, Manitoba, where polar bears have often foraged in recent years), you wander into town looking for something to eat, miserably out of sync with this brave new world. You may not realize it, but a developing hole in the Arctic ozone layer soon may give you skin cancer.

We have had nearly four decades to tally the worldwide toll of organochlorines since Rachel Carson's (1962b, 7) statement, in *Silent Spring*, that

The chemicals to which life is asked to make its adjustment are no longer merely the calcium and silica and copper and all the rest of the minerals washed out of the rocks and carried in rivers to the sea; they are the synthetic creations of man's inventive mind, brewed in his laboratories, and having no counterparts in nature.

For the first time in the history of the world, Carson (1962a, 24) wrote,

Every human being is now subjected to contact with dangerous chemicals, from the moment of conception until death. In the less than two decades of their use, the synthetic pesticides have been so thoroughly distributed throughout the animate and inanimate world that they occur virtually everywhere.

During the early twentieth century, human beings learned to synthesize chemicals for the first time on a scale and with a practicality that has since mightily shaped our commerce. For the first time, many thousands of synthetic substances were discharged into the environment with little forethought regarding how they might interact with the preexisting biological landscape. These chemicals are some of the most potent toxins ever developed. For example, one drop of dioxin, spread among 1,200 people, could kill all of them (Cook and Kaufman 1982, 1).

*The Dirty Dozen* begins in chapter 1 with a basic primer on synthetic organochlorines—what they are, how they originated, and why they are dangerous. Chapter 1 is organized around a historical narrative of synthetic chemistry, its uses, and its abuses, including a number of notable accidents after which scientists learned, often for the first time, what synthetic organochlorines can do to human and animal biology. Chapter 2 focuses on a half-dozen areas in which persistent organic pollutants have had a dramatic impact. This account begins in the Arctic with Inuit mothers who have learned they may be poisoning the next generation by feeding them POP-laced breast milk. Chapter 3 is on the chemical legacy that connects the depletion of stratospheric ozone with global warming. The interlocking chemistry of stratospheric ozone depletion and global warming also has implications for the Inuit, because the ozone layer has been rapidly thinning over their heads.

Chapter 4 introduces, through a number of local reports, the toll of synthetic toxins on minority ethnic groups in the United States. We begin with the Rosebud Sioux of South Dakota and continue to Akwesasne, land of the toxic turtles, a Mohawk reservation bisected by the U.S.–Canada border. The chapter concludes in "Cancer Alley," Louisiana, along the Mississippi River between Baton Rouge and New Orleans, with its ranks of petrochemical manufacturing plants.

Chapters 5, 6, and 7 of this book are devoted to a survey of ways in which persistent organic pollutants inflict physiological damage on humans and other animals. The reader should be prepared to enter a brave new chemical world in which vultures in India die from eating contaminated carrion, a world where being toxic may save the whales, at least for a few generations. It is a world of belugas with tumors and sterile bald eagles.

Chapter 8 of *The Dirty Dozen* focuses specifically on estrogenic effects that are creating a notable amount of sexual confusion. The reader may be startled to learn that, in a world of burgeoning humanity, sperm counts have declined as much as 50 percent in 50 years. On this journey, readers should be prepared to meet polar bears, frogs, and alligators with two sets of sexual organs. Such hormone-disrupting persistent contaminants can be hazardous at extremely low doses and pose a particular danger to offspring exposed in the womb. During prenatal life, endocrine disrupters can alter development and undermine the ability to learn, to fight off disease, and to reproduce.

Chapter 8 surveys public-policy solutions to the POP dilemma. An international ban of the dirty dozen was signed in Stockholm during December 2000, setting out control measures regarding the production, import, export, disposal and use of the twelve most notorious POPs. According to this treaty, governments are to promote the best available technologies and practices for replacing existing POPs while preventing the development of new ones.

Having negotiated an international ban (and one, unlike the Kyoto Protocol, which has the support of U.S. President George W. Bush), the struggle to eliminate POPs cannot yet be retired to a history of environmental finished business. The Stockholm Convention is a design for work to be done that will cause fundamental changes in many industrial processes from the printing of books to the manufacture of Barbie dolls.

Quite without malicious intentions, the chemical industry has put all of us in its experimental beaker. "We are conducting a vast toxicologic experiment, and we are using our children as the experimental animals," said Philip Landrigan, chairman of preventative medicine at the Mount Sinai School of Medicine (Moyers 2001).

According to Joe Thornton,

It makes no sense to presume that the thousands of organochlorines not yet tested will somehow turn out to be safe, when virtually every organochlorine investigated to date causes one or more toxic effects. . . . Addressing organochlorine as a class does not mean an outright ban. Because chlorine chemistry has invaded every corner of the economy, it would be disastrous to end production and use of organochlorines overnight. The only practical approach is a planned process to replace chlorine-based technologies gradually with cleaner alternatives. (Thornton 2000, 353, 355)

Synthetic organochlorines are created by combining the molecules of existing elements (usually, for the chemicals under discussion here, carbon, hydrogen, and chlorine) into combinations not heretofore present in nature. The manufacture of such chemicals accelerated during World War II, when, by accident, scientists seeking chemical weapons discovered that their creations killed insects. By the late 1940s, several insecticides (the most notable, at the time, being DDT) were introduced to combat insect infestations. The new chemicals were sprayed liberally on forests, farms, and towns. The production of synthetic pesticides rose fivefold during the fifteen years between 1947 and 1962, when *Silent Spring* was published. By the middle 1990s, 100,000 synthetic chemicals had been invented, and roughly 1,000 new ones were being introduced each year (Colborn, Dumanoski, and Myers 1996, 226).

The production of synthetic pesticides, herbicides, and fungicides for consumption in the United States began in earnest at the end of World War II and had risen to about 300 million kilograms a year by 1962, when Rachel Carson published *Silent Spring*. Production peaked in 1975 at slightly more than 700 million kilograms a year and then began a slow decline with the enforcement of environmental regulations (Pimentel and Lehman 1993, 94).

Synthetic organochlorines have become big business.

At least 11,000 organochlorines are produced commercially. Many thousands more are created as by-products during the manufacture, use and disposal of these initial 11,000. Only a few of the commercial organochlorines have been tested for carcinogenicity and a few other effects. Fewer still have been tested for their abilities to disrupt the human endocrine system, alter the sexual, neurological and immunological development during pre- and postnatal periods, or any of a long list of equally subtle but potentially devastating effects. Currently, it takes more than five years and hundreds of thousands of dollars to gather basic toxicological data for one single compound. (Allsopp, Costner, and Johnston 1995)

Expenditures on pesticides in the United States, in constant 1996 dollars, increased from roughly $3 billion in 1979 to about $10 billion in 1997 (Berenbaum, 2000, 59). By 1998, the global chemical pesticide market was $31 billion, a figure that was rising between 1 and 2 percent a year (Berenbaum, 2000, 145). The United States has no centralized reporting of pesticide expenditures, and so all figures are estimates. Expenditures appear to be rising roughly 3 percent per year in the United States, indicating a total of perhaps $11 billion by 2002.

Of all the pollutants released into the environment by human activity, POPs are among the most dangerous. They are highly toxic, causing an array of adverse effects, notably death, disease, and birth defects, among humans and animals. Specific effects can include cancer, allergies and hypersensitivity, damage to the central and peripheral nervous systems, reproductive disorders, and disruption of the immune system.

POPs have been implicated in the rising incidence of certain cancers (e.g. breast, prostate) and rising rates of endometriosis, as well as reproductive disorders such as infertility, declining sperm counts, fetal malformations, neurobehavioral impairment, and immune-system dysfunction.

In *Silent Spring,* Rachel Carson's main focus was DDT, which was invented by a German chemist in 1874. It was not used to kill insects until 1939, however. As an insecticide, DDT was first used on a large scale during the Naples typhus epidemic of 1943–1944; DDT was used by the Allied armed forces during the rest of World War II before the "war" against insects was joined on the home front in 1945.

Certain themes in *Silent Spring* resonate through the decades, even though the scientific landscape of environmental toxicology has changed materially in four decades:

I contend, therefore, that we have allowed these chemicals to be used with little or no advance investigation of their effects on soil, water, wildlife, and man himself. Future generations are unlikely to condone our lack of prudent concern for the integrity of the natural world that supports all life. (Carson 1962a, 13)

In less than two decades of their use, the synthetic pesticides have been so thoroughly distributed throughout the animate and inanimate world that they occur virtually everywhere. (Carson 1962a, 15)

[T]he chemist's ingenuity in devising insecticides has long ago outrun biological knowledge of the way these poisons affect the living organism [the human body]. (Carson 1962a, 25)

Agencies concerned with vector-borne disease are at present coping with their problems by switching from one insecticide to another as resistance develops. But this cannot go on indefinitely, despite the ingenuity of the chemists in supplying new materials. . . . [W]e are traveling a one-way street. No one knows how long that street is. If the dead end is reached before control of disease-carrying insects is achieved, our situation will indeed be critical. (Carson 1962a, 271)

Two present-day observers note that the register of potentially toxic organochlorines has multiplied since the time of *Silent Spring:*

Today, we can add vastly to Rachel Carson's list [of problematic chemical toxins]. Humankind is exposed to thousands of other chemical substances in ever-increasing quantity and variety. Of the 11 million substances known, some 60,000 to 70,000 are in regular use. Yet, toxicological data are only available for a fraction of the more than 3,000-odd chemicals that account for 90 percent by mass of the total used. The data on the environmental and ecotoxicological properties of such substances are even more scanty. (Van Emden and Peakall 1996, x)

Carson wrote only a decade and a half after organochlorines became widespread in the world environment. From the development of DDT grew a chemical industry that was still expanding as Carson described it in *Silent Spring*. Four decades later, Carson's original list of environmental troublemakers has multiplied several-fold. *Silent Spring* contains barely a mention of some chemicals that are at the top of everyone's list today, such as dioxins, PCBs, and chlorofluorocarbons (CFCs).

By the middle 1980s, new pesticides were being introduced by industry much more quickly than science and government could evaluate them for health hazards. The National Academy of Sciences estimated at that time that only 10 percent had been tested adequately to permit accurate assessment of health hazards; on 38 percent of pesticides available on the open market, for anyone's use, *no* toxicity information existed (Marco, Hollingsworth, and Durham 1987, 10).

These chemicals resist the natural processes of degradation by light, reactions with other chemicals, and biological processes. They are, therefore, potentially toxic for decades or centuries. The body cannot readily excrete persistent contaminants except through breast-feeding, and so most of the targeted POPs typically have long half-lives in the body, and with continued exposure their concentrations grow higher over time. Persistent contaminants are now pervasive in the food web, with animal products, including meat, fish, and milk, the primary routes of human exposure (World Wildlife Fund 2000).

No wonder even George W. Bush supports the Stockholm Convention to eliminate the dirty dozen. Evidence described in the chapters to come indicates just how dangerous these chemicals have become for the ecosystem of the entire Earth, especially for the most vulnerable people and animals in the Arctic. For them, POPs pose a fundamental threat to the survival of future generations.

## REFERENCES

Allsopp, Michelle, Pat Costner, and Paul Johnston. 1995. *Body of Evidence: The Effects of Chlorine on Human Health*. London: Greenpeace International. http://www.greenpeace.org/~uk/science/hdc/body.txt.

Allsopp, Michelle, Ben Erry, Ruth Stringer, Paul Johnston, and David Santillo. 2000. "Recipe for Disaster: A Review of Persistent Organic Pollutants in Food." Greenpeace Research Laboratories and University of Exeter (U.K.) Department of Biology. March. http://www.greenpeace.org/~toxics/reports/recipe.html.

Berenbaum, May, ed. Committee on the Future Role of Pesticides in U.S. Agriculture. Board on Agriculture and Natural Resources and Board on Environmental Studies and Toxicology. Commission on Life Sciences, National Research Council. *The Future Role of Pesticides in U.S. Agriculture.* Washington, D.C.: National Academy Press, 2000.

Carson, Rachel. 1962a. *Silent Spring.* Boston: Houghton-Mifflin.

Carson, Rachel. 1962b. *Silent Spring.* Westport, Conn.: Fawcett.

Colborn, T., D. Dumanoski, and J. P. Myers. 1996. *Our Stolen Future: Are We Threatening Our Fertility, Intelligence, and Survival? A Scientific Detective Story.* New York: Penguin.

Cook, Judith, and Chris Kaufman. 1982. *Portrait of a Poison: The 2,4,5-T Story.* London: Pluto Press.

Diamond, E. 1963. "The Myth of the 'Pesticide Menace.'" *Saturday Evening Post,* September 21, 17–18.

Marco, Gino J., Robert M. Hollingsworth, and William Durham, eds. 1987. *Silent Spring Revisited.* Washington, D.C.: American Chemical Society.

Moyers, Bill. 2001. "Trade Secrets: A Moyers Report." Program transcript. Public Broadcasting Service, March 26. http://www.pbs.org/tradesecrets/transcript.html.

Pimentel, David, and Hugh Lehman, eds. 1993. *The Pesticide Question: Environment, Economics, and Ethics.* New York: Chapman and Hall.

Thornton, Joe. 2000. *Pandora's Poison: Chlorine, Health, and a New Environmental Strategy.* Cambridge, Mass.: MIT Press.

Van Emden, Helmut F., and David B. Peakall. 1996. *Beyond Silent Spring: Integrated Pest Management and Chemical Safety.* London: Chapman and Hall and United Nations Educational Program.

Waddell, Craig, ed. 2000. *And No Birds Sing: Rhetorical Analysis of Rachel Carson's "Silent Spring."* Carbondale: Southern Illinois University Press.

Whitten, J. L. 1966. *That We May Live.* Princeton, N.J.: D. Van Nostrand.

World Wildlife Fund. 2000. "Toxics—What's New." http://www.worldwildlife.org/toxics/progareas/pop/pop_rep.htm.

# 1
# Persistent Organic Pollutants: The Basics

Commercial organochlorines have become pervasive in our everyday lives. They are as close at hand as the pesticides many people use on their lawns and gardens, the solvents auto mechanics use in their shops, and the plastics we use to store our foods. Organochlorines (*organo* refers here to the presence of carbon in combination with chlorine) are part of many lubricants and refrigerants. They are used in soaps, shampoos, deodorants, and cosmetics, as well as toothpastes and mouth rinses. Many children's toys, including the ubiquitous Barbie doll, are made of flexible polyvinyl chloride (PVC) (Allsopp, Costner, and Johnston 1995).

The "Dirty Dozen" of this book's title refer to the twelve chemicals that have been short-listed for elimination by international protocol. They include organochlorine pesticides (DDT, chlordane, mirex, hexachlorobenzene, endrin, aldrin, toxaphene, heptachlor) and industrial chemicals including PCBs, as well as the super-toxic dioxins and furans.

Persistent organic pollutants (POPs) are chlorine-based chemical compounds and mixtures that share four characteristics: high toxicity, persistence, a special affinity for fat, and a propensity to evaporate and travel long distances. Most people might be surprised to learn how common these toxic compounds are in their daily lives. Incinerators of all types emit dioxins, furans, and PCBs. Several times in Europe during the 1990s, emissions from municipal-waste or hazardous-waste incinerators have deposited these toxins on nearby grazing land, leading to contamination of milk. Persistent contaminants typical of industrial regions such as the Great Lakes have been found in albatrosses on remote Midway Island in the middle of the Pacific. Penguins in Antarctica have been found to be contaminated with a breakdown product of the pesticide chlordane and other persistent chemicals.

Each of these synthetic chemicals is built on a chlorine base.

Assessing the potential of any given chemical, either in the human body or in the environment, one question is of overriding importance: does it contain chlorine? Chlorine is highly reactive—that is, it combines very readily with certain other elements and it tends to bind to them very tightly. . . . Chlorine's ability to snap firmly into place, and to anchor all sorts of chemical structures, has made it, in the words of W. Joseph Stearns, director of chlorine issues for the Dow Chemical Company, "the single most important ingredient in modern [industrial] chemistry." (McGinn 2000, 33)

Heat-resistant PCBs became standard liquid insulation in electrical transformers, which are widely used in all electrical grids. Roughly 70 percent of all PCBs ever manufactured are still in use or in the environment, often in landfills, where they gradually seep into water tables (McGinn 2000, 34). Reclamation of copper from cables causes the release of dioxins, furans, and PCBs (from the burning of the PVC plastic coatings on the cables). In southern Taiwan, fish from aquaculture ponds have been found to be highly contaminated from the nearby burning of electrical cables and credit cards. Two accidental polyvinyl chloride fires in Germany in 1992 and 1996 also caused local contamination by dioxins and furans. For a time, the sale of several food products was banned in Germany. The use of chemical pesticides has been a central part of the "Green revolution," which aims to increase crop yields in Asia.

Brominated flame retardants, designed to reduce the fire hazard in personal computers and other electrical consumer products, have been found in whales living in the deep oceans. Synthetic musk compounds, used as fragrances in detergents, fabric softeners, and other household products, have been found in human breast milk. TributylTin (TBT), an additive in ship paint, is held responsible for the development of male sex characteristics in female snails. TBT is detectable in fish, mussels, and other marine life, particularly in harbors and coastal waters.

"These examples may just be the tip of the iceberg," said Paul Johnston, head of Greenpeace's research laboratories. "For substances which have been banned for some time, levels in wildlife are decreasing only slowly. At the same time, many chemicals with similarly hazardous properties continue to be used and released to the environment every day" (POPs Invade 1999).

Since 1945, annual global production of pesticides has increased twenty-seven-fold, from 0.1 million tons to 2.7 million tons (McGinn 2000, 32). In addition, pesticide formulations that are sold to the public have become more potent; current pesticide formulations often are ten to a hundred times as toxic as they were in 1975 (McGinn 2000, 32). About 11,000 organochlorines are in production. The most common plastic is now polyvinyl chloride (PVC). Global production of PVC rose more than 70 percent, from 12.8 million tons to 22 million tons, between 1988 and 1996 (McGinn 2000, 32).

Many new chemical products have been brought to market to increase human comfort and convenience (and corporate profits), with little forethought to their eventual impacts (two of which are carcinogenesis and the destruction of stratospheric ozone) on the natural world and its inhabitants. The number of chemicals registered with the U.S. Environmental Protection Agency (EPA) for commercial use reached 80,000 during 2000; roughly 3,800 of these are regarded by the agency as high-production-volume chemicals. Fewer than half these chemicals have been tested for toxicity to humans. Only 8 to 10 percent have been tested specifically for their effects on the developing brain, the developing immune system, the developing reproductive organs, or the endocrine system of babies (Moyers 2001). Tens of thousands of chemicals reached the market before testing was required.

The United Nations Environmental Program's Governing Council in 1995 identified twelve persistent organic pollutants as subjects of an eventual ban on manufacture and use worldwide because they damage the ecosphere and the diversity of life supported by it. While many of these pollutants had been banned in the United States, other countries continued to manufacture and use them until they were outlawed by international protocol in late 2000. The international ban of the dirty dozen (see the following list) was negotiated in Stockholm late in 2000; ratification of the protocol had been largely completed as this book was being prepared for publication.

**The Dirty Dozen**

Aldrin

> *Uses/Production*: Control termites and protect corn, potatoes, and cotton from insects
>
> *Treaty Action*: Elimination

Chlordane

> *Uses/Production*: Protect agricultural crops and control termites
>
> *Treaty Action*: Elimination

DDT

> *Uses/Production*: Protect agricultural crops and kill insects that carry malaria and typhus
>
> *Treaty Action*: Elimination; exemptions for disease control

Dieldrin

> *Uses/Production*: Control termites, soil insects, and insects carrying disease
>
> *Treaty Action*: Elimination

Dioxins and furans

> *Uses/Production*: By-products from waste incineration and industry
>
> *Treaty Action*: Reduction and minimization

Endrin

> *Uses/Production*: Protect cotton and grains from insects, rodents, and birds
>
> *Treaty Action*: Elimination

Heptachlor

> *Uses/Production*: Control termites, soil insects, and insects carrying malaria
>
> *Treaty Action*: Elimination

Hexachlorobenzene

> *Uses/Production*: Protect wheat from pests; also a by-product of some chemical manufacture
>
> *Treaty Action*: Elimination

Mirex

> *Uses/Production*: Control fire ants, leafcutter ants, and termites; fire retardant in plastics
>
> *Treaty Action*: Elimination

Polychlorinated biphenyls (PCBs)

> *Uses/Production*: Several industrial uses, mainly as electrical insulators
>
> *Treaty Action*: No new production; reduced use

Toxaphene

  *Uses/Production*: Protect agricultural crops and kill ticks and mites on livestock

  *Treaty Action*: Elimination

Several of these synthetic chemicals (including PCBs and dioxins, the best known) actually refer not to single chemicals, but to families of chlorine-based compounds. For example, the term *dioxin*, as used by toxicologists, refers to a number of substances that share certain molecular features. They contain chlorine atoms attached to carbon atoms that are linked to each other in structures known as aromatic rings. Polychlorinated biphenyls, for example—the notorious PCBs—often are considered dioxins in discussions of toxicity.

Many of the dirty dozen were first formulated to kill insects. Aldrin, one of the dirty dozen, is a pesticide used to control soil insects such as termites, corn rootworm, wireworms, weevils, and grasshoppers. It has been widely used to protect crops such as corn and potatoes and has been effective in protecting wooden structures from termites. Like many organochlorines, Aldrin is very toxic to human beings and other animals. A lethal dose of aldrin for an adult man (about five grams), provokes muscle twitchings, myoclonic jerks, and convulsions.

## THE PERSISTENCE AND MOBILITY OF POPS

Once a persistent contaminant has evaporated, it can travel thousands of miles in air masses, often hitchhiking on particles (such as dust) in the atmosphere. Through a process known as the "grasshopper effect," persistent chemicals jump around, evaporating in warm conditions and then settling in cool spots. When the temperature is right, POPs will again take flight and continue their hopscotching travels. Scientists have detected POPs everywhere on Earth they have looked for them, even in regions where these synthetic chemicals have never been used. For example, Eric Dewailly first visited the Canadian Arctic to establish a baseline for contamination in urban areas of Quebec. He quickly found POPs there at levels that were higher than those in Canadian cities (Dewailly et al. 1993, 173–76).

The pollution of PCBs (among other organochlorines) is truly global. They have been measured in ocean-water samples from the Sargasso Sea, the North Sea, the western North Pacific, the South Pacific, the Indian Ocean, and the seas around Antarctica, among other places thousands of miles from their emission sources (Thornton 2000, 34). Most POPs (most notably PCBs and dioxins) diffuse easily throughout the ecosphere. Many of these pollutants also flow (via ocean and air currents) to "cold sinks," such as the Arctic, the Antarctic, and higher elevations of mountain ranges.

The persistence of organochlorine POPs varies considerably; some have half-lives of less than a year, but others' decay rates are measured in millions (and even billions) of years. Hexachloroethane, for example, has a half-life in pure water of 1.8 billion years, and 1,2-dichloroethylene's half-life is estimated at 21 billion years (Thornton 2000, 33).

Since POPs are soluble in fats (in technical terms, they are lipophilic), the highest levels are usually found in fatty foods such as meat, fish, and dairy products. POPs can be detected, however, at lower concentrations in vegetables, fruits, and cereals. Many POPs reach their human hosts through foods that have been liberally doused with these compounds on their way to the table, usually to minimize insect damage. Farm workers who harvest these foods sometimes have been exposed at unhealthy levels. Food-borne persistent organochlorines are readily absorbed into the human body. By the late 1980s, 300,000 farm-worker poisonings a year were being reported in the United States, and the United Farm Workers were making a major issue of banning the most dangerous herbicides and pesticides.

Direct application of organochlorine pesticides leads to residues on crops. Contamination of the human food chain has also occurred when contaminated wastes have been mixed with livestock feedstuffs or directly with food intended for human consumption.

Once they have been absorbed, these unnatural chemicals evade the body's defenses. They cannot be shed using the body's processes for detoxifying and excreting other wastes that could become toxic. Instead, they accumulate in body fat, from which they are released most rapidly during periods when fat stores are drawn down, such as periods of hunger (including aggressive dieting) and lactation associated with pregnancy.

Throughout a woman's lifetime, the store of persistent contaminants increases in her body fat. By unfortunate coincidence, the demands of pregnancy and breast feeding draw down these fat reserves, and so a load of contaminants a mother has taken decades to accumulate passes to her baby in a very short time. Even worse, these hormone-disrupting contaminants hit the baby at the most vulnerable period of human development. The same goes for most other mammals.

For these reasons, and others, children who are breast-fed in cold sink areas, such as in the Arctic among the Inuit, are at particular risk. (The fact that the Inuit diet centers on animals, such as polar bears and seals, which also are accumulating these chemicals in their body fat, adds to the danger.)

## DISSENTERS TO THE MAJORITY PARADIGM

The behavior of organochlorines in the atmosphere and oceans was largely unknown to human science before 1985. Despite the recent arrival of a new scientific paradigm on the issue, a minority of scientists disagree, as with climate change. For example, Ames, Profet, and Gold (1990) have made a case that animals, including humans, employ natural defenses against synthetic chemicals such as dioxins. Comparing *indole carbinol* (found in broccoli) and alcohol with dioxin, Ames, Profet, and Gold conclude that "animals have a broad array of . . . defenses to combat the changing array of toxic chemicals in plant food (nature's pesticides) . . . [T]hese defenses are effective against both natural and synthetic toxins."

These researchers argue that the human body is "well-buffered" against toxins because much of its cellular material is quickly replaced, some of it within a few days. DNA also repairs itself, they argue. They also believe that "Plant antioxidants do not distinguish whether carcinogens are synthetic or natural in origin." While synergism between synthetic chemicals can multiply their toxicity, "This is also true of natural chemicals, which are by far the major source of chemicals in the diet." Similarly, this school of thought maintains that fat-soluble natural toxins "bioconcentrate" as effectively as their artificial cousins. A cooked steak contains hydrocarbons that mimic dioxin, they contend.

## BIOACCUMULATION

The key to the potency (and toxicity) of POPs is bioaccumulation (also sometimes called biomagnification). While studying Great

Lakes wildlife, Theo Colborn found that microscopic organisms pick up persistent pollutants in sediments. They, in turn, are consumed by zooplankton, which are consumed by mysids (small, shrimp-like creatures). A smelt may eat the mysids, and a lake trout may eat the smelt. By the time a herring gull (or a human being) eats the lake trout, its body may contain 25 million times the concentration of the pollutant found in a lake's sediment.

Another key to the potency of POPs is the tiny amount required to induce toxicity. Effects can be identified in doses measured in parts per billion—the equivalent, according to Colborn, et al., 1996) of a cocktail made from a few drops of gin and the tonic contained in a six-mile long train of railroad tank cars (Colborn, Dumanoski, and Myers 1996, 40). A single exposure of laboratory animals to some forms of dioxin (ranging from a dose of less than one microgram to a few milligrams per kilogram of body weight) is usually lethal (Lok and Powell 2000). Only short-term prenatal exposure of the mother to dioxins is required to produce long-term developmental and immunological problems in offspring during its entire lifespan (Lok and Powell 2000).

Even minute levels of PCBs, sometimes less than five parts per billion, may provoke biological damage over long periods of time. Monkeys fed a PCB level typical of concentrations found in human breast milk showed "significantly impaired learning and performance skills when tested at three years of age" (Schettler et al. 1999, 177). The monkeys' blood levels of PCB, at two to three parts per billion, were similar to typical levels in general human populations (Rice and Hayward 1997).

"The science is showing that extremely small amounts of dioxin delivered at critical times are enough to permanently disrupt fetal development and lead to very serious repercussions," said Frederick vom Saal, an expert on environmental chemicals at the University of Missouri at Columbia. "Today you don't find anybody saying small amounts of dioxin are likely to be safe," vom Saal said. "It's not a position that any responsible person is taking" (Allen 1997a).

## CONTAMINATION OF FOOD

A report issued by the Pesticide Action Network North America and Commonweal documents widespread contamination of U.S. food with many POPs that have been banned in the United States for many years. The report asserted that an average American may experience as many as seventy exposures a day to POPs, most of them in food. The report, "Nowhere to Hide: Persistent Toxic Chemicals in the U.S. Food Supply," analyzes chemical residue data collected by the Food and Drug Administration (FDA) and finds POPs in all food groups, from baked goods and meats to fresh fruits and vegetables (Schafer 2000).

These residues cannot be washed off produce with water. According to the report,

Virtually all food products are contaminated with POPs . . . including baked goods, fruit, vegetables, meat, poultry and dairy products. It is not unusual for daily diets to contain food items contaminated with three to seven POPs. A typical holiday dinner menu of 11 food items can deliver thirty-eight "hits" of exposure to POPs, where a "hit" is one persistent toxic chemical on one food item. . . . The top 10 POPs-contaminated food items, in alphabetical order, are: butter, cantaloupe, cucumbers (pickles), meatloaf, peanuts, popcorn, radishes, spinach, summer squash, and winter squash. (Schafer 2000)

A Greenpeace report published in March 2000 warned that much of the world's food is contaminated with POPs. The report, "Recipe for Disaster," reviewed existing data on food worldwide and revealed that some foodstuffs, particularly fish, meat, and dairy products, contain levels of POPs that exceed internationally agreed-upon limits (Allsopp et al. 2000). "Our food chain is the main route of human exposure to these highly toxic chemicals. The widespread contamination of food with manufactured chemicals in both industrialized and less-industrialized countries is fundamentally unacceptable," said David Santillo of the Greenpeace International Science Unit (Santillo 2000). "In many cases, we are being exposed to levels in excess of the maximum intakes deemed tolerable by international bodies," he added (Cooper 2000). Some POPs are produced intentionally, such as the pesticide DDT, which was first synthesized in Germany during 1874 but remained a laboratory curiosity until its pesticidal properties were discovered in 1939.

Allsopp and colleagues (2000) found that tolerable daily intakes for toxic POPs in food, as set by the World Health Organization (WHO), were exceeded in Spain, India, and parts of Canada as well as in sections of southern Sweden and southern Taiwan, where many people eat a large amount of fish. David Santillo also indicated that "Consumption of fish-oil dietary supplements may lead to excessive intake of dioxins in U[nited] K[ingdom] children" (Santillo 2000). The same report found that fish in Spain and Australia contained particularly high levels of certain POPs in excess of maximum levels set by the WHO. Meats in Vietnam and Mexico and dairy products in Hong Kong, Argentina, and Mexico were found to be similarly afflicted.

The same report (Santillo 2000) described instances in which food has been contaminated from local pollution sources, such as incinerators or metal recycling centers, as well as from the mixing of waste with animal feed, such as PCB-contaminated oils (as was the case during the 1999 dioxin chicken scandal in Belgium, which is described below). The report asserts that the scale and scope of POP pollution is not yet known, because there are large gaps in the scientific data on levels of POPs in food. Most studies are restricted to investigating a limited range of organochlorines, such as DDT and PCBs, while many other POPs remain unexamined. "Given that POPs are a global problem, it is likely that today's findings are just a glimpse at a much more widespread problem. Unless action is taken now to phase out the production and use of POPs, and the processes which generate them, our food and environment will remain contaminated for generations to come" (Santillo 2000).

Organochlorine levels in fish from the Great Lakes are still high, although levels have generally declined in recent years. DDT levels in meat have fallen in Australia and Canada. There have been some regional declines in PCBs in marine fish. DDT levels have generally decreased in marine fish but levels in areas of the tropics indicate continued input to the marine environment. In human milk, concentrations of dioxins and furans have not increased in Western countries in recent years, and levels in some European countries have declined. A decreasing trend has been observed for DDT in Europe, where it is banned, but not in Mexico, where it is still used. Levels of PCBs and chlordane in human milk do not appear to have declined in countries where levels have been monitored. These examples indicate that although levels of some compounds have declined, others remain stable because of their persistence and because contamination of the environment has continued.

Within the past few years, high levels of DDT have been detected in meat from Thailand, Vietnam, and Mexico. For dioxins and furans, veal is the most highly contaminated meat because of the high proportion of milk in calves' diets. In India, a high proportion of tested milk samples were found to be highly contaminated with DDT; some exceeded national limits. High levels of DDT were apparent in milk in Hong Kong and Argentina. Residues of aldrin and dieldrin were also reported to be high in milk from these countries, and levels of heptachlor and heptachlor epoxide exceeded safety standards in Hong Kong, Argentina, and Mexico.

DDT residues have been found in tea and coffee. A 1994 study reported that levels in vegetables from Australia gave evidence of recent application on crops despite a ban on using DDT there. Wheat stored in gunny sacks in India was found to be contaminated with DDT.

Sports anglers catching fish along the north shore of the Gulf of the St. Lawrence River and individuals eating fish from the Great Lakes carry notably high body burdens of organochlorines, including PCBs and dioxins (Dewailly et al. 1994; Fein et al. 1984). Elevated concentrations of dioxins and related organochlorines also have been documented in milk produced near municipal and industrial waste incinerators, as well as near other industries (Lassek, Jahr, and Mayer 1993; Liem et al. 1991).

In an attempt to protect public health, regulatory agencies have set permissible levels in specific foods, maximum residue limits, that should not be exceeded. For levels of organochlorines in diet, the Food and Agricultural Organization (FAO) and the WHO have set levels that are deemed to be "safe," called tolerable daily intakes (TDIs) or acceptable daily intakes (ADI). These regulatory limits are set using toxicity data from studies in laboratory animals and some data from human studies.

Risk assessments used to determine the ADIs/TDIs is a process that is fraught with uncertainties. Furthermore, results of toxicity testing in animals may be inappropriate for detecting certain health effects, potentially leaving health effects undetected. Risk assessment assumes a threshold dose of a chemical below which there are no health effects, a dose at which acceptable levels of exposure may be set. However, for some POPs, there may be no such threshold dose. Another complicating factor: Usually people are exposed to mixtures of chemicals. Sometimes two or more chemicals have effects in synthesis only. Therefore, it is unlikely that limits established for singular organochlorines in diet are truly protective of human health in the real world of multiple exposures.

## A BRIEF HISTORY OF POPS

During 1774 Swedish pharmacist Carl Wilhelm Scheele released a few drops of hydrochloric acid onto a piece of manganese dioxide. Within seconds, a greenish-yellow gas arose, marking the first scientific observation, quite by accident, of chlorine. At first, observers thought the gas was a compound of oxygen. Not until the nineteenth century did the English chemist Sir Humphrey Davy recognize chlorine as a chemical element. Davy gave the element its name after the Greek word *khloros,* anglicized to "chloric gas" or "chlorine," which became the seventeenth element on the periodic table.

By midcentury, scientists and manufacturers came to understand chlorine's powerful disinfectant qualities, which stem from its ability to bond with and destroy the outer surfaces of bacteria and viruses. Chlorine was first used as a disinfectant during 1846 to prevent the spread of child-bed fever in the maternity wards of Vienna General Hospital in Austria. Later, during the same century, chemists began to experiment with compounds of chlorine as they learned that it has an extraordinary ability to extend a chemical "bridge" between various elements and compounds that would not otherwise react with each other. By the late 1920s and 1930s, synthetic organic compounds were being invented and marketed with no evident concern for environmental effects.

The first synthetic organochlorines were created by accident during the late eighteenth century as by-products of the synthesis and use of elemental chlorine. At about the same time, Charles Darwin's interest in evolution led him to conduct a number of experiments dealing with plant reproduction. During these investigations, Darwin discovered that plant growth and reproduction are controlled by hormones that deliver chemical "signals" throughout any given plant. Auxin was isolated during 1928, followed by research aimed at using these chemicals to increase crop yields by speeding up the growth of plants.

PCBs are a combination of chlorine atoms with a molecule containing two joined hexagonal benzene rings, known as a biphenyl. Before they were detected throughout the environment, and before their toxic effects were known, PCBs were deemed a wonder chemical useful in just about any application in which fire suppression is an advantage, notably in the transmission of high-voltage electricity. Various PCBs were used to impart inflammability to wood and plastics, to preserve and protect rubber, and to water-seal stucco. They found many uses in varnishes, inks, and paints, as well as pesticides. About 1970, PCBs were used to create "carbonless" carbon paper.

By the mid-1930s, workers and some customers at Halowax Corporation (later part of Union Carbide) and General Electric (GE) were suffering outbreaks of chloracne—"small pimples with dark pigmentation of the exposed area, followed by blackheads and pustules" (Montague 1993). During 1936 three workers at Halowax Corporation died. The company hired Harvard University researchers to determine a cause. The researchers tested the air in several Halowax manufacturing plants and then designed experiments to replicate these levels and expose experimental rats to them. They reported that

The chlorinated diphenyl is certainly capable of doing harm in very low concentrations and is probably the most dangerous [of the chlorinated hydrocarbons studied]. . . . These experiments leave no doubt as to the possibility of systemic effects from the chlorinated naphthalenes and chlorinated diphenyls. (Drinker et al. 1937, 283)

The Swan Chemical Corp. (later a division of Monsanto) had begun manufacturing PCBs in commercial quantities during 1929. Eight years later, following the aforementioned experiments, the Harvard School of Public Health hosted a one-day meeting on systemic effects of certain chlorinated hydrocarbons including "chlorinated diphenyl" (an early name for PCBs). The meeting was attended by representatives from Monsanto, GE, the U.S. Public Health Service, and the Halowax Corporation, among others (Montague 1993). At this meeting F. R. Kaimer, an assistant manager of GE's Wireworks at York, Pennsylvania, said that one GE factory

had in the neighborhood of 50 to 60 men afflicted with various degrees of this acne about which you all know. Eight or ten of them were very severely afflicted—horrible specimens as far as their skin conditions was concerned. One man died and the diagnosis may have attributed his death to Halowax vapors, but we are not sure of that. (Drinker et al. 1937, 283)

As early as the 1930s, tests of laboratory animals revealed estrogenic properties in several synthetic organochlorines. Organochlorine estrogen mimicry by DDT was reported in chickens as early as 1950 (Dodds and Lawson 1938; Dodds, Goldberg, and Lawson 1938; Dodds et al. 1938). The Danish chemist Søren Jensen was the first, during the middle 1960s, to detect residues of PCBs throughout the environment; he found it even in hair samples taken from his wife and daughter. By the middle 1990s, a battery of tests costing about $2,000 could reveal to anyone the presence, in his or her body, of roughly 250 chemical contaminants, most of them in trace amounts. The tests could be carried out on anyone, anyplace on Earth, with similar results (Colborn, Dumanoski, and Myers 1996, 106).

Coincident with the introduction of DDT into the world ecosystem, the first warning sign appeared that some manufactured chemicals might spell serious trouble. In 1944 scientists found residues of DDT in human fat. Seven years later, another study brought disturbing news of DDT contamination in the milk of nursing mothers. During the early 1950s, naturalists observed thinning eggshells and rapidly declining populations of bald eagles and other birds. In 1962 Rachel Carson documented the growing burden of contamination in *Silent Spring,* which detailed the devastating impact of persistent pesticides on wildlife and warned about hazards to human health.

Even as environmental alarm bells began to ring, the use of synthetic chemical pesticides grew swiftly throughout the 1960s and 1970s. Japan experienced a thirty-three-fold increase in pesticide use between 1950 and 1974; little increase in the agricultural yield was associated with this lavish use of pesticides. Between 1945 and the late 1970s, pesticide use rose tenfold in the United States, while crop losses to pests doubled (Van Emden and Peakall, 1996, 65).

In the meantime, by 1969, research indicated that dioxins in 2,4,5-T cause birth defects in mice. By this time, cooking oil was found to have been contaminated by PCBs. The "Yusho" incident occurred in Japan in 1968 and the "Yu-Cheng" incident occurred in Taiwan in 1979, both affecting around 2,000 people. Increased mortality rates were recorded following the incidents, and a broad spectrum of adverse health effects were reported. Children exposed *in utero* were most severely affected.

Within a century and a half, according to the Chlorine Chemistry Council (CCC), "The chlorine industry [was] support[ing] nearly 2 million jobs with an annual payroll of more than $52 billion" in the United States alone. "In fact," asserted the CCC, "Almost 40 percent of U.S. jobs and income are in some way dependent on chlorine" (Chlorine Chemistry Council 2001). Furthermore, asserted the CCC, "Chlorine is irreplaceable in our economy. A ban on chlorine's use would cost U.S. consumers more than $90 billion per year for alternative products and process—with no guarantee of equivalent performance or quality" (Chlorine Chemistry Council 2001). By the year 2000, approximately 12 million tons of chlorine were being produced in North America.

According to the CCC, the largest volume of chlorine, about 35 percent, is used in the production of other chemicals, including many pharmaceuticals. Plastics consume more than 25 percent of the yearly output. A significant amount of chlorine, around 18 percent, is used to produce solvents for metalworking, dry cleaning, and electronics. Roughly 10 percent is used for pulp and paper bleaching. Chlorine is also used in drinking-water purification, as well as wastewater and swimming-pool disinfection (Chlorine Chemistry Council 2001).

## AGENT ORANGE IN VIETNAM

At the time of the first Earth Day, in the spring of 1970, the United States was pouring dioxin (as an active ingredient of Agent Orange) on the jungles of Vietnam, Laos, and Cambodia in an attempt to defoliate the jungles and deny insurgents places to hide from aerial bombing. The guerrillas were said to be "fish" in a "sea" of rural peoples that would be stripped bare by defoliants. Between 1962 and 1971, at least 12 percent of southern Vietnam's land area was doused liberally with nearly 18 million gallons of 2,3,7,8-tetra-chlorodibenzo-p-dioxin (TCDD), the most potent of dioxin's many varieties (Schecter et al. 2001, 435).

U.S. armed forces dropped more bombs (measured by weight) on Vietnam than it dropped in the entire Pacific Theater during World War II. By 1971, more than 600 pounds of bombs *per person* had been rained on Vietnam. Between 12 percent (U.S. figure) and 43 percent (National Liberation Front figure) of South Vietnam's land area was sprayed at least once with defoliants, usually Agent Orange (Johansen 1972, 4).

Eighteen million gallons of Agent Orange was sprayed over vast tracts of Southeast Asian forests between 1962 and 1971 in concentrations up to 1,000 times as potent as dioxin-based herbicides sold over the counter in the United States (McGinn 2000, 26).

Large areas of the countryside became unfit for human habitation during the war, and for several years thereafter. The population of

Saigon, now Ho Chi Minh City, increased tenfold between 1954 (when the war began with French intervention) and 1970, from 300,000 to 3 million people (Johansen 1972, 4).

Most of the herbicides, including Agent Orange, were applied to populated areas without prior hazard testing. Agent Orange was applied in large amounts, often haphazardly, by troops taking part in "Operation Ranch Hand," whose participants proclaimed that "Only we can prevent forests," a play on Smokey the Bear's slogan, "Only you can prevent forest fires" (Johansen 1972, 4). Samples collected in 1970 and 1973 documented elevated levels of TCDD in milk samples, as well as in fish and shrimp. Nursing mothers who had been heavy consumers of fish were found to have the highest levels in their blood (Schecter et al., 2001, 435).

Soon after spraying of Agent Orange and other herbicides began during the late 1960s, reports increased of deformed births in unusually large numbers. Areas sprayed with Agent Orange later reported very high incidences of certain birth defects: anencephaly (absence of all or parts of the brain), spina bifida (a malformed vertebral column), and hydrocephaly (swelling of the skull).

The Saigon newspaper *Tin Sang* published descriptions of "monster babies" born to mothers in areas that had been sprayed. The newspaper reported that one woman "reported that her newly pregnant daughter was caught in a chemical strike, and fainted, with blood coming out her mouth and nostrils, and later from her vulva. She was taken to a hospital where she was later delivered of a deformed fetus" (Johansen 1972, 4). The same day (October 26, 1969), *Dong Nai,* another Saigon newspaper, published a photograph of a still-born fetus with a duck-like face and an abnormally twisted stomach. A day later, the newspaper reported that a woman in the Tan An district who had been soaked with Agent Orange had given birth to a baby with two heads, three arms, and twenty fingers (Johansen 1972, 4). Many other similar accounts were published before the South Vietnamese government shut down the newspapers for "interfering with the war effort" (Johansen 1972, 4). The South Vietnam health ministry also began to classify accounts of deformed births as state secrets.

When these accounts were presented to U.S. Department of Defense officials, they were, at first, dismissed as unconfirmed enemy propaganda. When the accounts persisted and began to be more specific and numerous, the U.S. armed forces finally stopped using

Agent Orange. The U.S. Air Force later found a "significant and potentially meaningful" relationship between diabetes and bloodstream levels of dioxins in its ongoing study of people who worked with the defoliant Agent Orange during the Vietnam War (Brown 2000, A-14; Institute of Medicine 1994). Members of the U.S. armed services who were exposed to high levels of dioxins were found to be more prone to development of diabetes than those with low levels of exposure. People with the highest exposure levels developed diabetes most rapidly. While it once dismissed reports of cancers caused by Agent Orange as groundless, three decades later the U.S. Army was giving a special medallion—the Order of the Silver Rose— to soldiers who had been afflicted.

By mid-2001, the U.S. Department of Veterans Affairs was soliciting applications for compensation from Vietnam veterans with any of a large number of "presumptive disabilities": chloracne, Hodgkin's disease, multiple myeloma, non-Hodgkin's lymphoma, soft-tissue sarcoma, acute and subacute peripheral neuropathy, and prostate cancer ("Bulletin Board" 2001). The same request for claims asserted that diabetes mellitus soon would be included in its list of dioxin-induced pathologies.

At roughly the same time, a panel advising the EPA, well-stocked with industry representatives, was still arguing whether dioxin is carcinogenic in human beings. After ten years of work, during the summer of 2000, the EPA released a 3,000-plus page *Draft Dioxin Reassessment,* which concluded that "TCDD (and possibly other closely related structural analogs, such as the chlorinated didenzofurans) are carcinogenic to humans and can cause immune-system alterations; reproductive, developmental, and nervous system effects; endocrine disruption, altered lipid metabolism; liver damage; and skin lesions" (Schecter et al. 2001, 436). The EPA study confirmed many other studies that had linked TCDD and other forms of dioxin to "cancer and cancer mortality at relatively high levels in chemical workers and in toxicity studies" (Becher, Steindorf, and Flesch-Janys 1998; Fingerhut et al. 1991; Flesch-Janys et al. 1995; Flesch-Janys et al. 1998; National Toxicology Program 1998).

An Air Force study, conducted between 1982 and 2000, followed about 1,000 people who serviced or flew aircraft carrying Agent Orange. Their health was compared to a similar number who also served in Vietnam but had no known defoliant exposure. "The data do not prove dioxin causes diabetes, only that there appears to be a correlation between its level in the blood and the disease. There could be non-causal reasons for the association" (Brown 2000, A-14). People with diabetes may have higher levels of dioxins in their blood because their bodies get rid of the chemical more slowly than most, for example.

Men who sprayed Agent Orange in Vietnam between 1962 and 1971 were followed to determine whether exposure to dioxin affected their children. Nervous-system disorders were found to be widespread. Spontaneous abortions, birth defects, and developmental delays also were noted—paradoxically, men who received low doses of dioxin tended to give birth to more children with these problems than those who had been exposed at higher levels (Wolfe, Micalek, and Miner 1995).

Dioxin levels remained very high during the year 2000 in some areas of Vietnam that had been sprayed with Agent Orange more than three decades earlier (Schecter et al. 1992). Tests of people in the city of Bien Hoa (population 390,000), about twenty miles north of Ho Chi Minh City (formerly Saigon), in particular, evidenced dioxin readings 135 times higher than levels in Hanoi, which was not sprayed. Levels in Hanoi were measured at 2 to 3 parts per billion of TCDD (roughly background level in today's world), while blood levels of 271 p.p.b. were found in the blood of people living in Bien Hoa (Schecter et al. 2001, 435).

Bien Hoa was used during the war as a major U.S. military base and chemical depot where dioxin was transported, some of which leaked into the neighboring Bien Hung Lake, which is connected to the nearby Dong Nai River. This aquatic system became the conduit for the spread of dioxin toxicity among many people in the area. The highest readings were obtained from people who lived near the lake and had eaten fish from it regularly.

High levels of dioxin were reported not only in people who had lived in the area during the war, but also in people who had moved to Bien Hoa from uncontaminated areas many years after the war ended, illustrating the persistence of dioxin contamination. High levels also were reported in the bodies of people born in the area many years after the war ended and the chemical stores (but not toxic residues) were removed. Even the U.S. Department of Defense admitted that at least one underground storage tank had ruptured before 1971, spilling between 5,000 and 7,000 gallons of Agent Or-

ange into the local groundwater. The liberated dioxins thereby filtered into the water table and moved to the lake and the river,
attaching themselves to plants and animals along the way.

The research of Arnold O. Schecter (an environmental scientist at
the University of Texas School of Public Health) and colleagues
clearly indicated how residents of Bien Hoa—some of whom had not
been born during the war—continued to acquire contamination.
The dioxin first dumped in the area by U.S. armed forces was bioaccumulating up the food chain, from phytoplankton to zooplankton, and then to fish consumed by people:

Persons new to this region and children born after Agent Orange spraying
ended also had elevated TCDD levels. This TCDD uptake was recent and
occurred decades after spraying ended. We hypothesize that a major route
of current and past exposures is from the movement of dioxin from soil into
river sediment, then into fish, and from fish consumption into people.
(Schecter et al. 2001, 435)

"We have a public-health crisis for the people living in Bien Hoa
City," said Arnold Schecter, a leader in study of dioxin levels in Vietnam ("Dioxin Levels" 2001, 4). "We're seeing increasing dioxin levels
in people now compared to what I was seeing in the 1980s," Schecter
said. "I would regard this as an emergency" (Verrengia and Tran
2000).

Schecter, who began his Vietnam dioxin studies in 1984, served
in the U.S. Army Medical Corps during the Vietnam War. Schecter
has visited Vietnam sixteen times since then. He believes that the
extensive use of Agent Orange during the war has made Vietnam
today the world's number-one hot spot for dioxin contamination.
His studies at Bien Hoa are only one example of a plague of dioxin-
induced toxicity that still afflicts many people in Vietnam.

Thanh Xuan, a "peace village" near Hanoi, houses a hundred children who are retarded, some with stunted limbs or twisted spines.
Most arrive at the peace village unable to walk, speak, or read.
Across Vietnam, rates of birth defects, miscarriages, and other complications are still uncommonly high almost three decades after
spraying of Agent Orange ended during 1971. Many of the deformed
children in the peace village were born to parents who were sprayed
during the war.

"If I wasn't here, I don't know what I would do," lamented Nguyen
Kim Thoa, fifteen, sitting in her bedroom beneath a Britney Spears

poster. A reporter described Thoa's delicate features, "wrapped in a shroud of spongy skin tumors and charcoal splotches sprouting bristles" (Verrengia and Tran 2000). "I wasn't able to go to school at home," Thoa said. "The children always made fun of me. In their eyes, I was a freak. Here, I have friends and teachers who love me" (Verrengia and Tran 2000). Thoa's father served in the South Vietnamese Army between 1978 and 1980 along the Cambodian border, an area that was heavily sprayed during the war.

Hoang Dinh Cau, chairman of a national Vietnamese panel that investigates the war's ongoing health consequences, estimates that about 1 million Vietnamese people are afflicted with dioxin poisoning, including 150,000 children. Thirty years after the war some rice paddies that were abandoned after spraying have not been reclaimed, as "soaring forests with 1,000 different tree species shriveled, replaced by weedy meadows that livestock won't graze. Farmers call the new growth 'American grass'" (Verrengia and Tran 2000).

In a sparsely decorated bedroom of a two-story concrete building, Bui Dinh Bi recalls his days with communist forces in Quang Tri, in what was then called South Vietnam. During the early 1970s, after he was exposed to Agent Orange, Bui's skin lesions changed from mosquito-bite-like bumps to tumors that cover his body thirty years later. Bui and his wife had eight children. The first was stillborn, and then the next five died in infancy. Their two surviving children are mentally retarded (O'Neill 2000). Bui lives with twenty-nine other veterans and seventy children in Friendship Village, near Hanoi, one of about a dozen similar communities that the Vietnamese government has established for veterans and children afflicted with dioxin toxicity.

The *Washington Post* reported in April 2000 that "Canadian researchers have found high levels of dioxins in children that were born long after the spraying ceased in 1971. The lingering contamination is so severe in some areas that if they were in the United States, they would be declared Superfund sites, requiring an immediate cleanup effort" (O'Neill 2000).

Malformed children, such as the ones at Friendship Village, should be very carefully studied before conclusions are drawn linking dioxin to them, Schecter said. To date, the studies performed, according to Schecter, "have been hypothesis-generating, but not conclusive" (O'Neill 2000). Crippling birth defects seen across Vietnam should not be blamed wholesale on Agent Orange, and may be due to conditions such as polio, encephalitis, and cerebral palsy, Schecter said (O'Neill 2000).

Le Cao Dai, who worked with Schecter during his research trips, has no such reservations. He said that Agent Orange has become part of the food chain for the people of South Vietnam—affecting their water and food supply—and remains a problem today. The directors of Friendship Village say all the veterans were exposed and all the children's parents served in South Vietnam during the war (O'Neill 2000).

## DIOXIN SPILL AT SEVESO, ITALY (1976)

On July 10, 1976, a chemical reaction went out of control and provoked an explosion at the Industrie Chimiche Meda Societa Arionaria (Icmesa) trichlorophenol manufacturing plant in the northern Italian town of Seveso, pumping clouds of dioxins from the subsidiary of the chemical giant Hoffman La Roche.

One source observed that "The local people knew of the explosion [at Seveso] because of the noise, followed by a reddish-brown cloud with a nasty smell" (Cook and Kaufman 1982, 10). Other accounts say the dioxin resembled snow so much that children rushed out of doors to play in it. Birds at Seveso, according to another eyewitness account, "literally fell out of the sky and off branches, contorted in death" (Cook and Kaufman 1982, 11). Domestic cats were found "swollen and half paralyzed. . . . Dogs vomited green froth and developed skin eruptions that looked like burns." The following early autumn, "two women who had been in the crucial [exposure] Zone A, nearest to the explosion, gave birth to babies with virtually no brain tissue." Both children died (Cook and Kaufman 1982, 11).

The plant's accidental plume consisted of nearly pure dioxin. Four days after the dioxin release, flowers and other plants began to wither and die in Seveso. Birds, fish, and other animals were reported to be dying in abnormally large numbers. Eight days after the dioxin release, word went out from Seveso seeking medical doctors from all points of the compass as the town's hospitals overflowed with hundreds of ailing people. Their dioxin-dusted homes had become uninhabitable. Many pregnant women were faced with having deformed children, or seeking an abortion against the wishes of the Catholic Church (Hynes 1989, 13).

The 1976 dioxin spill near Seveso, Italy, resulted in the highest
levels of the chemical ever recorded in humans. Its aftermath produced the first studies to confirm that elevated dioxin levels persisted in people from the exposed areas almost twenty years after
the accident. The same study also found that women experienced
higher dioxin levels than men, portending harm to mothers' offspring.

## LOVE CANAL BECOMES A TOXICOLOGICAL
## HOUSEHOLD WORD

During the 1940s, Hooker Chemical Corporation used the Love
Canal area (north of Buffalo, New York, along a body of water upstream from Niagara Falls) as a dumping ground for a variety of toxic
wastes amounting to 20,000 tons, some of which contained dioxin
contamination. During the early 1950s the contaminated area was
covered with a clay cap that was supposed to seal the contaminants
underground. The land was sold by Hooker Chemical, and, over the
next few years, became the site of several hundred family homes
and a school.

During construction, the clay cap was breached, at the same time
that highway construction disrupted water flow around the site.
Shortly thereafter, several homeowners noted a foul mixture of toxins oozing into their basements. After direct appeals to government
agencies brought little help, the residents of Love Canal organized
public protests that attracted considerable media attention.

In 1978 President Jimmy Carter declared Love Canal a federal
disaster area, the first such declaration for a human-induced environmental disaster. The government proposed to buy the 200
houses nearest the dump site, but residents were not happy with
the proposed settlement. During 1980, residents took two Environmental Protection Agency officials hostage for several hours. The
number of homes to be purchased was then raised to about 700. In
the meantime, Love Canal became a toxicological household word.

## TIMES BEACH, MISSOURI, CEASES TO EXIST

During the late 1960s and early 1970s, waste hauler Russell Bliss
was hired to oil the dusty roads of Times Beach, Missouri, twenty-
five miles southwest of St. Louis, off Interstate 44. Bliss unknow-

ingly mixed dioxin-contaminated waste with the oil that he used for dust control on several sites, including some of the roads and stables in Times Beach and twenty-six other eastern Missouri sites that later were identified as dioxin "hot spots" by the EPA. Bliss used waste from several companies, including a Verona, Missouri, plant that had been purchased by Agribusiness Technologies' parent company, Syntex Agribusiness. The plant made hexachlorophene, an antibacterial agent.

During 1982, the EPA took random soil samples at various sites in Missouri to test for dioxin levels and found, late in the year, what the agency considered to be dangerous levels of dioxin in the soil at Times Beach. The 2,240 citizens of Times Beach found themselves sitting on one of the most toxic patches of earth in the United States.

In February 1983 the EPA announced a buyout of the entire town, for almost $33 million. The federal government permanently closed the city after it was discovered that dioxin had contaminated the town as a by-product of chemical processing and incineration, as well as Bliss' spraying. The town was listed as a Superfund site and demolished in 1992. Cleanup of the site was finished in 1997.

During 1990 the state of Missouri, the EPA, and Agribusiness Technologies agreed to incinerate dioxin-contaminated material at Times Beach. The decision called for incineration at Times Beach of dioxin-contaminated soils from several sites. Times Beach had been a ghost town since 1983, when it was purchased using Superfund monies. Syntex was held responsible and estimated the work would take five to seven years. Syntex also agreed to pay the government $10 million to reimburse some of the government's costs in the case.

The decision to incinerate Times Beach's dioxin-laced soil sparked opposition, and six years of delays, from some environmental groups. On March 17, 1996, the incinerator began operation; by mid-June 1997, 265,354 tons of soil and other dioxin-contaminated material from Times Beach and twenty-six other sites in eastern Missouri had been incinerated. David Shorr, director of the Missouri Department of Natural Resources, said that the cost of the Times Beach cleanup was about $200 million (Mansur 1997).

The incineration may have removed dioxin from the soil of what used to be Times Beach. Taking a planetwide inventory, however, the dioxin was merely displaced, given up to the winds and waters, and lodging, perhaps, eventually in the milk of an Inuit mother, or that of a polar bear. Pat Costner, a Greenpeace scientist, said that neither the EPA nor its state counterparts denied that some un-known quantity of dioxin was dispersed by the incinerator into the environment.

"That's contrary to protecting the public's health or the environ-ment," Costner said. "Now we'll just have to watch and wonder for decades" about the effect of the dioxin dispersal, Costner said. "That's what is left now" (Mansur 1997).

Within two months after Times Beach shut down its incinerator, dioxin was discovered in the soil in west St. Louis County—this time in the suburb of Ellisville. Officials with the EPA said that a private driveway in Ellisville had been found to have dioxin levels as high as 195 parts per billion. In a statement, the agency said dioxin levels more than 1 part per billion are "of concern" (Allen 1997b).

## A SUPERFUND SITE ON THE HUDSON

In 1983, the same year that Times Beach became uninhabitable, two hundred miles of the Hudson River were declared a Superfund hazardous-waste site. The EPA ordered GE to dredge thirty years' worth of PCBs from a thirty-five-mile stretch of the river. The com-pany had discharged PCBs into the Hudson from manufacturing plants in Fort Edward and Hudson Falls between 1947 and 1977. Between Albany and Fort Edward, two GE capacitor plants dumped more than 1 million pounds of PCBs into the Hudson River before PCBs were outlawed in 1977.

The EPA in early August 2001 ordered GE to pay $480 million to dredge toxic chemicals, mainly PCBs, off the floor of the upper Hud-son River. This was the costliest environmental cleanup in the United States to that time.

The anticipated cleanup will last more than five years and involve "dredging 2.65 million cubic yards of contaminated sediment along a picturesque 37-mile run of the Hudson River" (Powell 2001, A-5).

Dredging of the Hudson was being planned with machinery that sucks up contaminated sludge and water. One account described the scope of the project:

The dredging, slated to start in 2003, will be epic in sweep. Dredgers will ply the river's waters, at least two processing plants will be built on the river or its banks to separate the PCBs, and a stream of trains and trucks will haul thousands of pounds of the foul-smelling stuff to presumably distant landfills. Once that is completed, the dredgers will lay down thousands more tons of clean sediment on the riverbed to cover remaining PCBs and begin restoration of the natural habitat. (Powell 2001, A-5)

Before the EPA order, GE officials spent at least $15 million on an advertising and lobbying campaign against dredging. GE officials asserted that the safer course, not to mention the less expensive one, would be to let natural sediments inexorably entomb the PCBs on the riverbed. The river is vastly cleaner than it was three decades ago, and GE officials note that thousands now swim in the river and at least one town in the affected region—Waterford, New York—taps the river for drinking water (Powell 2001, A-5).

The company spent millions of dollars on lobbying and advertising, in an abortive attempt to persuade state and federal officials that the dredging project, which will remove roughly 150,000 pounds of PCB-laced sediments from the river's bed, is unnecessary and environmentally risky (Lazaroff 2001).

During August 2001, GE said that massive dredging would "cause more harm than good" (Lazaroff 2001). "This is a loss for the people of the area who overwhelmingly oppose this project and the decades of disruption it will bring to their communities," GE said in August. "It appears that neither sound science nor the voices of these residents played a part in the EPA's decision" (Lazaroff 2001). GE has argued that the EPA's cleanup plan would "resuspend" PCBs now buried in river sediments, thereby releasing additional toxins into the environment. In sticking by its original plan, the EPA has rejected that argument (Lazaroff 2001).

According to plans presented during December 2001, the design phase of the dredging project is expected to take about thirty-six months, and the EPA may adjust the plan after reviewing data collected during this phase. As part of the process, the EPA will solicit public input on issues including the location for the initial dredging and the type of equipment to be used. The dredging itself could take at least five years (Lazaroff 2001). During April 2002 GE said it would comply with the EPA order.

## OPPOSITION TO POPS GALVANIZES

Even before an explosion of DDT use in the late 1940s, scientists at the U.S. Fish and Wildlife Research Center in Laurel, Maryland, published papers establishing its toxicity to wildlife from sprays at high rates of application (5 pounds per acre) (Marco, Hollingsworth, and Durham 1987, 85–87). The nature of this toxicity remained unknown to the general public until publication of Rachel Carson's *Silent Spring* in 1962. Pest resistance to DDT also became public knowledge with the publication of *Silent Spring,* but knowledge of resistance in Italian houseflies was noted in small-circulation scientific journals as early as 1948 (Marco, Hollingsworth, and Durham 1987, 120). Similarly, the persistent nature of certain synthetics (notably DDT and PCBs) was known to science by the middle 1960s, long before public knowledge forced their restriction, regulation, and eventual elimination.

During 1987, the Conservation Foundation (Washington, D.C.) and the Institute for Research on Public Policy (Ottawa, Ontario) convened a team of scientists to study the ecological condition of the Great Lakes and their watersheds. This study included effects of synthetic chemicals on wildlife. A book, *Great Lakes/Great Legacy?* (1990) was based on the proceedings of this conference; it asserted that "chemical contamination was associated with the widespread reproductive failure of wildlife within the watershed, particularly that of fish and fish-eating birds and mammals" (Krimsky 1999, 16).

At about the same time, Theo Colborn, who had compiled a literature search for the Great Lakes/Great Legacy project, was beginning to draw together the scientific threads that would establish the study of synthetic toxins' endocrinal effects. Colborn convened several conferences that brought together experts in biology, biochemistry, toxicology, and zoology, as she began to weave information from diverse sources into a pattern that linked incomplete sexual development of fish in the Great Lakes with increasing wildlife sterility, thyroid abnormalities, and declining rates of survival in several bird species in the same watershed. "From these observations," wrote Sheldon Krimsky, "she advanced a generalized causal hypothesis, suggesting that the pervasive presence of chemical endocrine disruptors introduced into the environment was responsible for the reproductive abnormalities found in wildlife populations" (Krimsky 1999, 20).

The first of a series of meetings convened by Colborn during July 1991, with financial assistance from the W. Alton Jones Foundation, was titled "Chemically Induced Alterations in Sexual and

Functional Development: The Wildlife/Human Connection." Known to its participants as "Wingspread I" the conference ended with a consensus statement signed by twenty-one participants that asserted that a large number of human-manufactured chemicals had been injected into the environment, that these chemicals are bioaccumulative across generations and along food chains, and that these chemicals, by mimicking estrogen, were interfering with the functioning of many species' endocrine systems.

Colborn recalls being surprised at what seemed, at the time, the consummate implausibility of the connections she and the others were uncovering, wondering whether many people would believe what seemed like the plot of a science-fiction novel. The Wingspread I consensus statement went on to list several specific ways in which endocrine-disrupting chemicals have been shown to disrupt reproduction in several species of animals.

## THE CHLORINE CHEMISTRY COUNCIL IS CREATED

The Chlorine Chemistry Council, created in 1993 by the Chemical Manufacturers Association, advocates use of chlorine compounds in industry and agriculture. It maintains that dietary elements (dairy products and soya) are causing increasing estrogen loads in people's bodies more than synthetic industrial chemicals. According to the CCC (known colloquially as "c-cubed") human intake of natural estrogens is "several orders of magnitude" above the contribution of synthetic toxins (Krimsky 1999, 101).

Started during the mid-1990s, the Chemical Industry Institute of Technology, "a respected research institute that is heavily supported by major chemical companies," (Krimsky 1999, 164) was spending about $1.5 million annually investigating the environmental endocrine hypothesis.

## HOW TO POISON A NATION'S FOOD WEB: THE CASE OF BELGIUM, 1999

An example of how easily one accident can imperil the food web of an entire nation was provided by Belgium during 1999 as animal feeds become contaminated with PCB-laced oils. As a result of this contamination, chickens were contaminated at levels up to 5,000 times safety levels. Not all the chickens were for human consumption; some were used for breeding chicks, contaminating egg supplies as well.

Belgium quarantined about a thousand poultry and pig farms during the summer of 1999, following detection of dioxin contamination, after tests revealed PCBs, which can indicate the presence of dioxins in animal feed. Dioxin was introduced into the Belgian food supply through contaminated animal fat used in animal feeds supplied to Belgian, French, and Dutch farms. Hens, pigs, and cattle ate the contaminated feed.

The feed was supplied by the Versele company, which used animal fat from Verkest, one of two fat processors identified as the source of a dioxin contamination episode in Belgium the previous May. Fat processor Fogra also was implicated in the original contamination. Dioxin contamination prompted bans on sales of Belgian meat, chicken, eggs, dairy products, and processed foods in several other countries.

The Belgian government's initial reluctance to reveal the severity of the contamination caused political upheaval that led voters to defeat the incumbent coalition government in elections held during June 1999. The crisis reportedly cost the Belgian economy more than $1.5 billion in lost revenue (Handyside 1999). The government of Belgium was forced to repurchase and destroy pork and processed pork foods with a fat content of more than 20 percent that may have been contaminated.

Investigations later revealed that Belgian authorities had known of the contamination since mid-March; the high dioxin levels had been confirmed on April 26, a month before the government made the information public. Only on May 28, 1999, did the Belgian public health minister announce a ban on the sale of dioxin-laced chickens and eggs.

Contaminated fat had been used to manufacture animal feed by Verkest as early as late January 1999. The contaminated feed was supplied to eight other Belgian animal-feed manufacturers, as well as one each in France and the Netherlands. The same feed was then sold to egg, broiler chicken, pork, and beef producers.

On May 26, test results revealed high levels of dioxins in laying hens at farms receiving feed from the implicated animal-feed producers, indicating that chickens and eggs sold during April had contained high levels of dioxins. Beginning on May 27, the 417 poultry farms that had bought feed from these nine suppliers were placed under surveillance. Their products were traced. On May 28 the Belgian public health minister ordered removal of all chicken and eggs from store shelves and cautioned the public against eating Belgian poultry and eggs (Lok and Powell 2000).

The initial ban was extended by the government on June 1 to include a ban on sale of all products containing chicken or eggs until they were inspected. Slaughter and transport of cattle, pigs, and poultry were prohibited. A preliminary list of all poultry and later pork and beef farms using feed originating directly or indirectly from Verkest was assembled and updated on a regular schedule. As the number of banned foods grew, the agriculture and public health ministers of the Belgian government resigned on June 1 (Lok and Powell 2000). At least thirty countries (including Canada, Australia, Hong Kong, Taiwan, Russia, Egypt, Algeria, South Africa, Poland, Switzerland, most European Union countries, and several other non–European Union countries) temporarily banned imports of Belgian agricultural products (Lok and Powell 2000).

On June 2, the European Union directed member nations to remove and destroy all animal feed, poultry, and egg products that might contain dioxins, as well as "other products containing more than 2 percent egg product produced by the suspect farms between January 15 and June 1" (Lok and Powell 2000). Throughout June, countries across Europe quit selling all Belgian egg and poultry products. The same day, two Verkest executives, father and son, were arrested and charged with fraud and falsification of documents that prosecutors said "misled customers [into] thinking they were buying 100 percent animal fat" (Lok and Powell 2000). The accused denied that they had deliberately contaminated the suspect animal feed.

The European Community on June 4 extended its restrictions to include products derived from livestock other than poultry (including milk and dairy products) from suspect farms. On June 10, angry Belgian farmers blockaded roads at the Dutch and French borders in protest against the export bans.

Belgian officials arrested two executives from Fogra on June 23. A public prosecutor said Fogra was the initial source of contamination, having dioxins and PCBs in its products and suggested that motor oil there had been mixed with frying fat. On July 14, investigators said they were almost sure that oil from an electrical transformer somehow had found its way into old cooking oil that was recycled into fat for animal feed (Lok and Powell 2000).

As problems related to tainted animal feed spread across Europe, media reports provided Belgians with discomforting detailed reports describing in graphic detail the process by which animal fat is procured for feed. Reporters found that fat-processing plants routinely collected used fat from municipal recycling facilities where consumers disposed of used motor oils along with other household wastes. The plants also accept used fat and oil from restaurants and industrial food plants, which filter, cook, and sterilize it. The recycled fat was then shaped into blocks and sold to animal-feed manufacturers. Some reports speculated that dioxin-tainted transformer oil was illegally dumped into cooking oil somewhere in this process.

On June 30, the Belgian government said that such problems had cost the nation's economy US$1.54 billion, half from the agricultural sector and the other half from other food industry ("Belgium Sees" 1999). On July 23, 233 more pig farms were quarantined after two of them reported PCB levels up to fifty times allowable limits. Several Belgian scientists wrote an article in the scientific journal *Nature* contending that, given usual consumption patterns, most people who ate food produced in Belgium would not suffer PCB-related harm (Bernard et al. 1999, 231).

Dioxin contamination during these years was not limited to Belgium. Contaminated citrus pulp pellets from Brazil in animal feed caused dioxin contamination of milk in Germany during late 1997 and early 1998. Some milk had levels greater than national safety limits. The citrus pulp also was used as animal feed in eleven other European countries. About 92,000 tons of citrus pulp pellets had to be thrown away. The contamination arose from dioxin-contaminated waste lime produced as a by-product by the Solvay company in Brazil. The waste lime is converted into a form that is then added to citrus pulp for animal feed.

## LOCAL PROTESTS RISE

Protests of POPs' toll may erupt anyplace with a significant point-source of the chemicals, usually from a factory's smokestacks or waterborne wastes. Incinerators are favorite targets. For example, during the summer of 2001 activists stepped up their protests at an East Oakland, California, medical waste incinerator, blocking its access roads until they were arrested and carried away by police. Local environmentalists and community activists have been protesting at Integrated Environmental Systems, the state's only commercial medical-waste incinerator, claiming it emits cancer-causing dioxin and other hazardous chemicals into the air. Incinerator officials deny their operations pose a health risk to the community, contending that their emissions are well below state and federal standards.

"Yesterday," wrote a reporter for the *San Francisco Chronicle* on August 7, 2001, "was the first time protesters attempted to disrupt incinerator operations by keeping trucks carrying waste from entering or leaving the facility, just off Interstate 880 on High Street. The group, which numbered about 70 people before dwindling to fewer than 40, did allow a few trucks to pass through, but said there were far fewer than normal" (DeFao 2001, A-16). Several protesters said they live within a mile of the plant and fear for their health.

"Today is the beginning of the end of incineration," said Bradley Angel of Greenaction in San Francisco, one of the protest leaders. "If the government won't do it, the people will" (DeFao 2001, A-16).

## REFERENCES

Allen, W. 1997a. "Dioxin Find Worries Residents: Many Have Questions about Chemical." *St. Louis Post-Dispatch,* August 10. http://lists. essential.org/1997/dioxin-l/msg00271.html.

———. 1997b. "Dioxin Levels Found in Private Drive in Ellisville." *St. Louis Post-Dispatch,* August 6. http://lists.essential.org/1997/dioxin-l/msg 00271.html.

Allsopp, Michelle, Pat Costner, and Paul Johnston. 1995. *Body of Evidence: The Effects of Chlorine on Human Health.* London: Greenpeace International. http://www.greenpeace.org/~toxics/reports/recipe.html.

Allsopp, Michelle, Ben Erry, Ruth Stringer, Paul Johnston, and David Santillo. 2000. "Recipe for Disaster: A Review of Persistent Organic Pollutants in Food." Exeter, U.K.: Greenpeace Research Laboratories and University of Exeter Department of Biology. March. http://www.greenpeace.org/~toxics/reports/recipe.html. (This report includes a lengthy bibliography of scientific literature on the subject.)

Ames, Bruce N., Margie Profet, and Lois Swirsky Gold. 1990. "Nature's Chemicals and Synthetic Chemicals: Comparative Toxicology." http://www.mapcruzin.com/environment21.

Becher, H., K. Steindorf, and D. Flesch-Janys. 1998. "Quantitative Cancer Risk Assessment for Dioxin Using an Occupational Cohort." *Environmental Health Perspectives* 106: 663–70.

"Belgium Sees Dioxin Crisis Costing 60 Billion Belgian Francs." 1999. Reuters News Service, June 30.

Bernard, A., C. Hermans, F. Broeckaert, G. De Poorter, A. De Cock, and G. Houins. 1999. "Food Contamination by PCBs and Dioxins: An Isolated Episode in Belgium Is Unlikely to Have Affected Public Health." *Nature* 401: 231–34.

Brown, David. 2000. "Defoliant Connected to Diabetes." *Washington Post,* March 29. http://irptc.unep.ch/pops/newlayout/press_items.htm.

"Bulletin Board: Vietnam Veterans Benefit from Agent Orange Rules." 2001. *Indian Country Today,* May 16.

Chlorine Chemistry Council. 2001 (May). Home Page. http://www.c3.org/chlorine_what_is_it/chlorine_story2.html.

Colborn, T., D. Dumanoski, and J.P. Myers. 1996. *Our Stolen Future: Are We Threatening Our Fertility, Intelligence, and Survival? A Scientific Detective Story.* New York: Penguin.

Cook, Judith, and Chris Kaufman. 1982. *Portrait of a Poison: The 2,4,5-T Story.* London: Pluto Press.

Cooper, Jack. 2000. "World Food Supplies Contaminated with Toxic Chemicals." Greenpeace International. March 19. http://greenpeace.org/~toxics/reports/recipe.html.

Costner, Pat. 2000. *Dioxin Elimination: A Global Imperative.* Amsterdam, Neth.: Greenpeace International. www.greenpeace.org/~toxics under "reports." [Can be also found at http://www.chej.org.]

DeFao, Janine. 2001. "Protesters Block Waste Operation; Three Arrested at Medical Incinerator." *San Francisco Chronicle,* August 8.

Dewailly, E., P. Ayotte, S. Bruneau, C. Laliberte, D. C. G. Muir, and R. J. Nordstrom. 1993. "Human Exposure to Polychlorinated Biphenyls through the Aquatic Food Chain in the Arctic." *Organohalogen Compounds* 14: 173–76.

Dewailly E., J. J. Ryan, C. Laliberte, S. Bruneau, J. P. Weber, S. Gingras, and G. Carrier. 1994. "Exposure of Remote Maritime Populations to Coplanar PCBs." *Environmental Health Perspectives* 102, suppl. 1: 205–9.

"Dioxin Levels Still Up in Many Vietnamese." 2001. *Omaha World-Herald,* May 15.

Dodds, E. C., L. Goldberg, and W. Lawson. 1938. "Oestrogenic Activity of Esters of Diethyl Stilboestrol." *Nature* 142: 211–12.

Dodds, E. C., L. Goldberg, W. Lawson, and R. Robinson. 1938. "Oestrogenic Activity of Alkylated Stilboestrols." *Nature* 141: 247–48.

Dodds, E. C., and W. Lawson. 1938. "Molecular Structure in Relation to Oestrogenic Activity: Compounds without a Phenanthrene Nucleus." *Proceedings of the Royal Society of London* 125, suppl. B: 222–32.

Drinker, Cecil K., et al. 1937. "The Problem of Possible Systemic Effects from Certain Chlorinated Hydrocarbons." *Journal of Industrial Hygiene and Toxicology* 19 (September): 283–311.

Fein, G. G., J. L. Jacobson, S. W. Jacobson, P. M. Schwartz, and J. K. Dowler. 1984. "Prenatal Exposure to Polychlorinated Biphenyls: Effects on Birth Size and Gestational Age." *Journal of Pediatrics* 105: 315–20.

Fingerhut, M. A., W. E. Halperin, D. A. Marlow, L. A. Piacitelli, P. A. Honchar, M. H. Sweeney, A. L. Greife, P. A. Dill, K. Steenland, and A. J. Suruda. 1991. "Cancer Mortality in Workers Exposed to 2,3,7,8-tetrachlorodibenzo-p-dioxin." *New England Journal of Medicine* 324: 212–18.

Flesch-Janys, D., et al. 1995. "Exposure to Polychlorinated Dioxins and Furans (PCDD/F) and Mortality in a Cohort of Workers from a Herbicide-Producing Plant in Hamburg, Federal Republic of Germany." *American Journal of Epidemiology* 142: 1165–75.

Flesch-Janys, D., J. Steindorf, P. Gurn, and H. Becher. 1998. "Estimation of the Cumulated Exposure to Polychlorinated dibenzo-p-dioxins/furans and Standardized Mortality Ratio Analysis of Cancer Mortality by Dose in an Occupationally Exposed Cohort." *Environmental Health Perspectives* 106: 655–62.

Handyside, Gillian. 1999. "New Dioxin Food Scare Strikes Belgium." Reuters News Service. August 4. http://platon.ee.duth.gr/data/maillist-archives/oikologia/1998-9/msg00337.html.

Huff, J. 1994. "Dioxins and Mammalian Carcogenesis." In *Dioxins and Health,* edited by A. J. Schecter. New York: Plenum.

Hynes, H. Patricia. 1989. *The Recurring Silent Spring.* New York: Pergamon Press.

Johansen, Bruce. 1972. "Ecomania at Home; Ecocide Abroad." *University of Washington Daily,* May 24.

Kamrin, Michael A., and Paul W. Ridgers. 1985. *Dioxins in the Environment.* Washington, D.C.: Hemisphere Publishing.

Kazman, Sam, Eric Askanase, and Julie DeFalco. 1996. "A Petition to Declare Times Beach, Missouri, a National Historic Landmark." http://www.cei.org/MonoReader.asp?ID = 518.

Krimsky, Sheldon. 1999. *Hormonal Chaos: The Scientific and Social Origins of the Environmental Endocrine Hypothesis.* Baltimore, Md.: Johns Hopkins University Press.

Lassek, E., D. Jahr, and R. Mayer. 1993. "Polychlorinated Dibenzo-p-dioxins and Dibenzofurans in Cow's Milk from Bavaria." *Chemosphere* 27, no. 4: 519–34.

Lazaroff, Cat. 2001. "EPA Authorizes Hudson River Cleanup." Environment News Service, December 5. http://ens-news.com/ens/dec2001/2001L-12-05-06.html.

Liem, A. K. D., R. Hoogerbrugge, P. R. Koostra, E. G. Van der Velde, and A. P. J. M. De Jong. 1991. "Occurrence of Dioxins in Cow's Milk in the Vicinity of Municipal Waste Incinerators and a Metal Reclamation Plant in the Netherlands." *Chemosphere* 23: 1675–84.

Lok, Corie, and Douglas Powell. 2000. "The Belgian Dioxin Crisis of the Summer of 1999: A Case Study in Crisis Communications and Management." February 1, 2000. http://www.plant.uoguelph.ca/risk comm/crisis/belgian-dioxin-crisis-feb01-00.

Mansur, Michael. 1997. "After 15 Years, Dioxin Incineration at Times Beach, Mo. Is Finished." *Kansas City Star,* June 18. http://archive.nandotimes.com/newsroom/ntn/health/061897/health1_8068.html.

Marco, Gino J., Robert M. Hollingsworth, and William Durham, eds. 1987. *Silent Spring Revisited.* Washington, D.C.: American Chemical Society.

McGinn, Anne Platt. 2000. "POPs Culture." *World Watch,* March–April, 26–36.

Montague, Peter. 1993. "How We Got Here, Part 1; The History of Chlorinated Diphenyl (PCBs)." *Rachel's Environment and Health News* 327, March 4. http://www.rachel.org/bulletin/bulletin.cfm?Issue_ID = 802 &bulletin_ID = 48.

Moyers, Bill. 2001. "Trade Secrets: A Moyers Report." Program transcript. Public Broadcasting Service, March 26. http://www.pbs.org/trade secrets/transcript.html.

National Toxicology Program. 1998. *Report on Carcinogenesis: TCDD.* Bethesda, Md.: National Institute of Environmental Health.

O'Neill, Annie. 2000. "Damaged Lives: Vietnamese Veterans and Children: While World Leaders Debate the Effects of Agent Orange, a Multinational Project Reaches Out to People at the Center of the Storm." *Washington Post* article reprinted in *Pittsburgh Post-Gazette,* November 5. http://groups.yahoo.com/group/VeteranIssues/message/364.

"POPs Invade Far Reaches of the Earth." 1999. Environment News Service, August 12. http://ens.lycos.com/ens/aug99/1999L-08-12-04.html.

Powell, Michael. 2001. "EPA Orders Record PCB Cleanup; GE to Foot $480 Million Bill to Dredge Upper Hudson River." *Washington Post,* August 2.

Rice, D. C., and S. Hayward. 1997. "Effects of Postnatal Exposure to a PCB Mixture in Monkeys on Non-spatial Discrimination Reversal and Delayed Alternation Performance." *Neurotoxicology* 18, no. 2: 479–94.

Santillo, David. 2000. "World Chemical Supplies Contaminated with Toxic Chemicals." Greenpeace Listserve, March 19. http://www.green peace.org/~toxics/reports/recipe.html.

Schafer, Kristin. 2000. "Nowhere to Hide: Persistent Toxic Chemicals in the U.S. Food Supply." Press release, Pesticide Action Network North America., November. http://www.panna.org/panna/resources/resources.html.

Schecter A., ed. *Dioxins and Health.* New York: Plenum, 1994.

Schecter, Arnold, O. Papke, M. Ball, D. C. Hoang, C. D. Le, Q. M. Nguyen, T. Q. Hoang, N. P. Nguyen, H. P. Pham, K. C. Huynh, D. Vo, J. D. Constable, and J. Spencer. 1992. "Dioxin and Dibenzofuran Levels in Blood and Adipose Tissue of Vietnamese from Various Locations in Vietnam in Proximity to Agent Orange Spraying." *Chemosphere* 25: 1123–28.

Schettler, Ted, Gina Solomon, Maria Valenti, and Anne Huddle. 1999. *Generations at Risk: Reproductive Health and the Environment.* Cambridge, Mass.: MIT Press.

Thornton, Joe. 2000. *Pandora's Poison: Chlorine, Health, and a New Environmental Strategy.* Cambridge, Mass.: MIT Press.

U.S. Environmental Protection Agency. 1990. "Times Beach Settlement Reached." Press release, July 20. http://www.epa.gov/history/topics/times/01.htm.

Van Emden, Helmut F., and David B. Peakall. 1996. *Beyond Silent Spring: Integrated Pest Management and Chemical Safety.* London: Chapman and Hall and United Nations Educational Program.

Verrengia, Joseph B., and Tini Tran. 2000. "Vietnam's Children Feeling Effects of Agent Orange." *Amarillo* [Texas] *Globe-News,* November 20. http://www.amarillonet.com/stories/112000/hea_agentorange.shtml.

Wolfe, W. H., J. E. Michalek, and J. C. Miner. 1995. "Paternal Serum Dioxin and Reproductive Outcomes among Veterans of Operation Ranch Hand." *Epidemiology* 6, no. 1: 17–22.

# 2

# "We Feel like an Endangered Species": Toxics in the Arctic

"As we put our babies to our breasts we are feeding them a noxious, toxic cocktail," said Sheila Watt-Cloutier, a grandmother who also is Canadian president of the Inuit Circumpolar Conference (ICC). "When women have to think twice about breast-feeding their babies, surely that must be a wake-up call to the world" (Johansen 2000, 27). Watt-Cloutier was raised in an Inuit community in remote northern Quebec. She didn't know it at the time, but toxic chemicals were being absorbed by her body and by those of other Inuit in the Arctic.

Watt-Cloutier now ranges between her home in Iqaluit (pronounced "ee-ha-loo-eet," capital of the new semisovereign Nunavut Territory) to and from Ottawa, Montreal, New York City, and other points south, doing her best to alert the world to toxic poisoning and other perils faced by her people. The ICC represents the interests of roughly 140,000 Inuit spread around the North Pole from Nunavut (which means "our home" in the Inuktitut language) to Alaska and Russia. Nunavut itself, a territory four times the size of France, has a population of roughly 27,000, 85 percent of whom are Inuit.

Many residents of the temperate zones hold fond stereotypes of a pristine Arctic largely devoid of the human pollution that is so ubiquitous in their lives. To a tourist with no interest in environmental toxicology, the Inuit Arctic homeland may seem as pristine as ever during its long, snow-swept winters. Many Inuit still guide dogsleds onto the pack ice surrounding their Arctic-island homelands to hunt polar bears and seals. Such a scene may seem pristine, until one realizes that the polar bears' and seals' body fats are laced with dioxins and PCBs.

The toxicological due bills for modern industry at the lower latitudes are being left on the Inuit table in Nunavut, in the Canadian
Arctic. Native people whose diets consist largely of sea animals
(whales, polar bears, fish, and seals) have been consuming concentrated toxic chemicals. Abnormally high levels of dioxins and other
industrial chemicals are being detected in Inuit mothers' breast
milk.

To the naked eye, the Arctic still *looks* pristine. In Inuit country
these days, however, it's what you *can't* see that may kill you:

• Persistent organic pollutants ("POPs" to environmental toxicologists),
such as DDT, dioxins, and PCBs, are multiplying up the food chain of the
Inuit as air and ocean currents transport the effluvia of southern industry
into polar regions. Geographically, the Arctic could not be in a worse position for toxic pollution, as a ring of industry in Russia, Europe, and North
America pours pollutants northward. POPs have been linked to cancer,
birth defects, and other neurological, reproductive, and immune-system
damage in people and animals. At high concentrations, these chemicals also
damage the central nervous system. Many of them also act as endocrine
disrupters, causing deformities in sex organs as well as long-term dysfunction of reproductive systems. POPs also can interfere with the function of
the brain and endocrine system by penetrating the placental barrier and
scrambling the instructions of the naturally produced chemical messengers. The latter tell a fetus how to develop in the womb and postnatally
through puberty; should interference occur, immune, nervous, and reproductive systems may not develop as programmed by the genes inherited by
the embryo.

• Global warming is accelerating more quickly in the Arctic than anywhere else on Earth; some Inuit villages are being washed into the sea or
slowly swallowed by melting permafrost as shrinking sea ice imperils the
survival of some species. Sand flies have been seen on Banks Island in the
high Arctic. Several Inuit have fallen through thin ice, usually on snow-
mobiles, and died. Residents of Banks Island, above the Arctic Circle, tell
of experiencing thunder, lightning, and hail for the first time in anyone's
memory.

• Atmospheric chemistry in the stratosphere is accelerating ozone depletion over the Arctic (see also chapter 3), threatening the Inuit and animals
with cancer from ultraviolet radiation. Some elders and hunters in Iqaluit
have reported physical abnormalities afflicting the seals they catch, including some seals without hair, "and seals and walruses with burn-like holes
in their skin" (Lamb n.d.).

Welcome to ground zero on the road to environmental apocalypse:
a place, and a people, who never asked for any of the curses that

industrial societies to the south have brought to them. The bevy of environmental threats facing the Inuit are entirely outside their historical experience.

"We are the miner's canary," says Watt-Cloutier. "It is only a matter of time until everybody will be poisoned by the pollutants that we are creating in this world" (Lamb n.d.). "At times," said Watt-Cloutier, "We feel like an endangered species. Our resilience and Inuit spirit and of course the wisdom of this great land that we work so hard to protect gives us back the energy to keep going" (Sheila Watt-Cloutier, personal communication, March 28, 2001).

Most the chemicals that now afflict the Inuit are synthetic compounds of chlorine; some of them are incredibly toxic. For example, one-millionth of a gram of dioxin will kill a guinea pig (Cadbury 1997, 184). Scan the list of scientific research funding around the world and add up what ails the Arctic. In addition to a plethora of studies documenting the spread of persistent organic pollutants through the flora and fauna of the Arctic, many studies aim to document the saturation of the same area by levels of mercury, lead, and nuclear radiation in fish and game (PD 2000 Projects 2001).

Following is a sampling of titles for research projects directed at documenting the ongoing eco-tastrophe in the Arctic. The destruction of Arctic's ecosystem is certainly being well documented. The studies have been generated by scientists in the United States, Canada, Sweden, Norway, and Russia (PD 2000 Projects 2001).

"New Persistent Chemicals in the Arctic Environment"

"Retrospective Survey of Organochlorines and Mercury in Arctic Seabird Eggs"

"Effects of Prenatal Exposure to Organochlorines and Mercury in the Immune System of Inuit Infants"

"Estimation of Site Specific Dietary Exposure to Contaminants in Two Inuit Communities"

"Follow-up of Pre-school-aged Children Exposed to PCBs and Mercury through Fish and Marine Mammal Consumption"

"Temporal Trends of Persistent Organic Pollutants and Metals in Ringed Seals of the Canadian Arctic"

"Assessment of Organochlorines and Metal Levels in Canadian Arctic Fox"

"Effects and Trends of POPs on Polar Bears"

"Contaminants in Greenland Human Diet"

"Lead Contamination of Greenland Birds"

"UV-Radiation and Its Impact on Genetic Diversity, Population Structure, and Foodwebs of Arctic Freshwater"

"Effects of Metals and POPs on Marine Fish Species"

"Heavy Metals in Grouse Species"

"Endocrine Disruption in Arctic Marine Mammals"

"Persistent Toxic Substances, Food Security, and Indigenous Peoples of the Russian North"

"Metals in Reindeer"

"Concentrations and Patterns of Persistent Organochlorine Contaminants in Beluga Whale Blubber"

"Ultraviolet (UV) Monitoring in the Alaskan Arctic"

The bodies of some Inuit, thousands of miles from the sources of the pollution, have the highest levels of PCBs ever detected, except in victims of industrial accidents. Some native people in Greenland have more than seventy times as much of the pesticide hexacloro-benzene (HCB) in their bodies as temperate-zone Canadians (Johansen 2000, 27).

Pesticide residues in the Arctic today may include some used decades ago in the southern United States. The Arctic's cold climate also slows decomposition of these toxins, and so they persist in the Arctic environment longer than at lower latitudes. The Arctic acts as a cold trap, collecting and maintaining a wide range of industrial pollutants, from PCBs to toxaphene, chlordane to mercury, according to the Canadian Polar Commission (PCB Working Group, n.d.). As a result, "Many Inuit have levels of PCBs, DDTs and other persistent organic pollutants in their blood and fatty tissues that are five to ten times greater than the national average in Canada or the United States" (PCB Working Group, n.d.).

Generation of POPs has become an issue in Watt-Cloutier's home, Iqaluit, on Baffin Island, where the town dump burns wastes that emit dioxins and furans. The dump's plume provides only a small fraction of Iqaluit residents' POP exposure, but it has become enough of an issue to figure in a three-month shutdown of the dump that caused garbage to pile up in the town. The dump was reopened after local public-health authorities warned that the backlogged garbage could spread disease and that "the hazard posed by the rotting piles of garbage outweighed the risks of burning it" (Hill

2001, 5). In 2001 Iqaluit's government was asking residents to separate plastics and metals from garbage that can be burned without adding POPs to the atmosphere.

Dioxin, PCPs, and other toxins accumulate with each succeeding generation in breast-feeding mammals, including the Inuit and many of their food sources. The reproductive cycle of humans and other mammals compounds the toxic effects of these chemicals. Airborne toxic substances are absorbed by plankton and small fish, which are then eaten by dolphins, whales, and other large animals. The mammals' thick subcutaneous fat stores the hazardous substances, which are transmitted to offspring through breast-feeding. Sea mammals are more vulnerable to this kind of toxicity than land animals, and so the levels of chemicals in their bodies can reach exceptionally high levels. The level of these toxins increases with each breast-fed generation.

## INUIT INFANTS "A LIVING TEST TUBE FOR IMMUNOLOGISTS"

Inuit infants have provided "a living test tube for immunologists" (Cone 1996, A1). Due to their diet of contaminated sea animals and fish, Inuit women's breast milk contains six to seven times the PCB level of women in urban Quebec, according to Quebec government statistics. Their babies have experienced strikingly high rates of meningitis, bronchitis, pneumonia, and other infections compared with other Canadians. One Inuit child out of every four has chronic hearing loss due to infections.

"In our studies, there was a marked increase in the incidence of infectious disease among breast-fed babies exposed to a high concentration of contaminants," said Eric Dewailly, a Quebec Public Health Center researcher and Laval University scientist who was among the first scientists to detect high levels of POPs in the bodies of Inuit mothers and their breast milk (Cone 1996, A1). Born with depleted white blood cells, Inuit children suffer excessive bouts of diseases, including a twentyfold increase in life-threatening meningitis compared to other Canadian children. These children's immune systems sometimes fail to produce enough antibodies to resist even the usual childhood diseases. A study published September 12, 1996, in the *New England Journal of Medicine* (Jacobson and Jacobson 1996, 783–89) confirmed that children exposed to low levels of PCBs in the womb grow up with low IQs, poor reading comprehension, difficulty paying attention, and memory problems.

According to the Quebec Health Center, a concentration of 1,052 parts per billion (ppb) of PCBs has been found in Arctic women's milk fat. This compares to a reading of 7,002 ppb in polar bear fat, 1,002 ppb in whale blubber, 527 ppb in seal blubber, and 152 ppb in fish. The U.S. Environmental Protection Agency safety standard for edible poultry, by contrast, is 3 ppb and in fish, 2 ppb. At 50 ppb, soil is often considered to be hazardous waste.

Research by the Canadian federal department of Indian and Northern Affairs indicates that Inuit women throughout Nunavut have DDT levels nine times that of averages in Canadian urban areas. Milk of Inuit women of the Eastern Arctic has been measured to contain as much as 1,210 ppb of DDT and its derivative, DDE, while milk from women living in southern Canada contains about 170 ppb (Suzuki 2000).

Although the United States is thousands of kilometers from its borders, Nunavut receives up to 82 percent of its dioxins from industries there (Suzuki 2000). Some of the pollution in the Canadian Arctic arrives from as far away as Western Europe and Japan. Contaminants can reach the Canadian Arctic from Europe in two weeks.

Dewailly accidentally discovered that Inuit infants were being heavily contaminated by PCBs. During the middle 1980s, Dewailly first visited the Inuit seeking a pristine group to use as a baseline with which to compare women in southern Quebec who had PCBs in their breast milk. Instead, Dewailly found that Inuit mothers' PCB levels were several times as high as the levels of the Quebec mothers in his study group.

Dewailly and colleagues (Dewailly, Ayotte, et al. 2000; Dewailly, Bruneau, et al. 1993; Dewailly, Dodin, et al. 1993; Dewailly, Ryan, et al. 1994) then investigated whether organochlorine exposure is associated with the incidence of infectious diseases in Inuit infants from Nunavut. Dewailly and colleagues (Dewailly, Bruneau, et al. 1993, 403–6) reported that serious ear infections were twice as common among Inuit babies whose mothers had higher than usual concentrations of toxic chemicals in their breast milk. More than 80 percent of the 118 babies studied in various Nunavut communities had at least one serious ear infection in the first year of life (Calamai 2000).

The most common contaminants that researchers found in Inuit mothers' breast milk were three pesticides (dieldrin, mirex, and

DDE, a derivative of DDT) and two industrial chemicals—PCBs and hexachlorobenzene. The researchers could not pinpoint which specific chemicals were responsible for making the Inuit babies more vulnerable because chemicals' effects may intensify in combination.

The Arctic Monitoring and Assessment Program, a joint activity of Arctic nations and organizations of indigenous Arctic peoples, said that "PCB blood levels, while highest in Greenland and the eastern Canadian Arctic, were high enough (over 4 micrograms of PCBs per liter of blood) that a proportion of the population would be in a risk range for fetal and childhood development problems" (PCB Working Group n.d.).

Because they are not easily broken down or excreted, these compounds remain in the body for months or years. In ecosystems, they tend to concentrate or bioaccumulate in the bodies of animals at the top of their respective food chains: large meat eaters such as marine mammals, polar bears, raptors, and human beings. Dolphins, seals, and whales in the northern seas are being contaminated. Large land animals, such as caribou, also are affected.

"The last thing we need at this time is worry about the very country food that nourishes us, spiritually and emotionally, poisoning us," Watt-Cloutier said. "This is not just about contaminants on our plate. This is a whole way of being, a whole cultural heritage that is at stake here for us" (Mofina 2000, A12). "The process of hunting and fishing, followed by the sharing of food—the communal partaking of animals—is a time-honored ritual that binds us together and links us with our ancestors," said Watt-Cloutier (PCB Working Group, n.d.).

In many cases, the Inuit have no practical alternative to "country food." Although grocery stores do business in Canadian Inuit hamlets today, all the food is flown in. No roads or natural land bridges lead south from the villages. The cost of air freight, compounded by distance, raises the cost of a quart of milk in the high Arctic to $4, and that of a battered head of lettuce to $3. A tiny frozen turkey the size of a stewing hen may cost $40 in the Arctic.

## TOXIC POLLUTION VIA THE DEW LINE

At the same time that Inuit are being poisoned by imported toxins, they also have discovered that parts of their homelands are laced with toxic "hot spots" left behind by abandoned military installations and mines, all also imports from the industrial south. Several of these hot spots are located at or near the 63 military sites in Canada, Greenland, and Alaska that make up the Distant Early Warning (DEW) system of radar sites. At these sites, according to the Arctic Monitoring and Assessment Program, an estimated thirty tons of PCBs were used, and "an unknown amount has ended up in their landfills" (PCB Working group, n.d.).

Under an agreement reached in 1998, Canadian taxpayers, not the U.S. government, are paying most of the $720 million cleanup bill for fifty-one decommissioned U.S. military sites across Canada. Cleanup of cancer-causing PCBs, mercury, lead, radioactive materials, and various petroleum by-products is expected to take nearly thirty years. Under the arrangement, the United States was absolved of legal responsibility for environmental damage in Canada in exchange for $150 million in U.S. weapons and other military equipment.

The cleanup of all fifty-one American military sites has revealed pollution that newspaper reports in Canada characterized as "staggering" (Pugliese 2001, B1). For example, at Argentia, Newfoundland, seventy miles southwest of St. John's, a large U.S. Navy base, which opened in 1941 and closed in 1994, left behind PCBs, heavy metals, and asbestos as well as landfills full of other hazardous wastes. Waste fuels also have contaminated the water table in the area.

Abandoned DEW sites in the Arctic were contaminated with discarded batteries, antifreeze agents, solvents, paint thinners, PCBs, and lead. According to news accounts, "[Canadian] Defence Department scientists have established that PCBs have leaked from the DEW line sites into surrounding areas as far away as 20 kilometers and, in some cases, the chemicals have been absorbed by plant and animal life" (Pugliese 2001, B1). Many of the DEW line locations were established in areas where native people hunt and fish.

Alaska Community Action on Toxics works with indigenous communities who face toxic contamination from Cold War sites, including the Yup'ik community on Saint Lawrence Island. Alaska Community Action on Toxics also has provided the first comprehensive map of more than 2,000 hazardous waste sites in Alaska.

The Inuit also endure pollution from the European and Asian side of the Arctic Ocean. Pesticide residues and other pollutants spill into the Arctic Ocean from north-flowing Russian and Siberian rivers. Decaying Russian nuclear submarine installations on the White Sea have polluted the ocean with nuclear waste, including entire reactor cores from scrapped ships.

While many of the former Soviet Union's worst polluters have gone out of business, some prosper despite the fact that their effluent is adding to the Canadian Arctic's toxic overload. The worst offender is the Norilsk nickel smelter, in northern Siberia. Traces of heavy metals from Norilsk's industries have been detected in the breast milk of Inuit mothers.

Geoffrey York, a reporter for the *Toronto Globe and Mail*, described the industrial city of Norilsk, population 230,000, as "the world's most polluted Arctic metropolis" (York 2001, A1).

Looming at the end of the road is a horizon of massive smokestacks, leaking pipes, rusting metal, gigantic slag heaps, drifting smog, and thousands of denuded trees as lifeless as blackened matchsticks. Inside malodorous smelters, Russian workers wear respirators as they trudge through the hot suffocating air, heavy with clouds of dust and gases. . . . Soviet[era] research in 1988 found that Norilsk Nickel had created a 200-kilometer corridor of dead forests to the southeast of the city. (York 2001, A11)

"We, the Inuit, who for a millennia have lived in harmony with, and with great respect for our land and wildlife, are now most impacted by outside forces such as POPs and climate change," said Watt-Cloutier. "With so much already on our plate in terms of attempting to reclaim and restructure our lives to gain back some control over our lives, be it personal, family, institutions, governance systems, etc., it can at times be overwhelming" (personal communication, Mar. 28, 2001).

## SOCIOECONOMIC CHANGES IN NUNAVUT

Nunavut could be characterized as the largest "Indian reservation" in the world. The per capita annual income in Nunavut is about U.S. $7,000, which doesn't go far in a place where two liters of milk costs $7 and a loaf of Wonder bread retails for $3. While most of Nunavut has few roads and no railroads (communities are connected by regular airline service), all villages now have television and radio, and most have telephone and the Internet, and so images imported from the outside world are never far away.

Widespread poverty, unemployment, crime, substance abuse, and a high suicide rate continue to beset Nunavut, which also faces 22 percent unemployment. Iqaluit (Inuktitut for "fishing place"), the capital of Nunavut (situated on southeastern Baffin Island about sixteen hundred miles north of New York City) contains about sixty-five hundred people, many of whom share big-city problems imported from the outside, including alcoholism, drug abuse, and AIDS.

The Royal Canadian Mounted Police detachment at Iqaluit has complained that its members spend so much time rounding up drunks that "it's getting in the way of fighting crime" (Iqaluit Drunks 2001).Three thousand incarcerations in the local drunk tank have been clocked in a year.

The traditional (usually nomadic) Inuit hunting economy endured largely intact until the 1950s and 1960s, when Canadian authorities forced many people to settle in permanent communities. Many Inuit children were pressured to attend church-run boarding schools away from their families. The schools were designed to assimilate Inuit children into the culture of the *qablunaaq*, the white man.

The near-collapse of the hunting economy has coincided with a sharp rise in suicides. Less than fifty years ago, suicide was virtually unknown among the Inuit of Canada's Arctic. Suicide now has become one of the leading causes of death in Iqaluit. Among young people, suicide has become by far the leading cause of death. Geela Patterson, adviser to the new territorial government on women's issues, and herself a suicide-prevention counselor, took her own life during June 1999. Even in a town with one of the highest suicide rates in Canada (seven times the Canadian average), where few families have been untouched, Patterson's death sent shock waves through the community.

Most of Nunavut's suicides in recent years have been healthy young men, aged fifteen to twenty-nine. After the destruction of fur markets from antifur campaigns in Europe and the United States, unemployment rates in northern villages have risen more than 50 percent. "When you reduce the usefulness of men in society, it is bound to have psychological effects," said Simona Arnatsiaq, leader of a Baffin Island women's organization (Brooke 2000).

Susan Aglukark, an Inuit singer who has gained pop-star status in southern Canada, "sings of the pain caused by suicide in ballads in English and in her Native Inuktitut" (Brooke 2000). "We are investing more money for suicide prevention," said Paul Okalik, whose brother committed suicide (Brooke 2000). "I myself, when younger and miserable, have heard the siren call of suicide as a release from suffering, as has my husband, and others that I know and love," Rachel Attituq Qitsualik, wrote in the *Nunatsiaq News*, a weekly newspaper serving the eastern Arctic. "My brother gave in to it" (Brooke 2000).

## GLOBAL WARMING: THE ARCTIC WORLD TURNED UPSIDE DOWN

Global warming is no theory in the Arctic, no subject of debate in search of (in U.S. President George W. Bush's words) "sound science." In many areas of the Arctic, unusual warmth in a land that lives and dies by snow and ice has become a daily companion that is turning a long-stable world upside down. Global warming and pollution are increasing in the Arctic north of Canada "at a much faster rate than even the most pessimistic predictions of a decade ago" (Bourrie 1998). Inland Arctic areas are warming by 1.5°C per decade ("Conservation" 1999). In 1999 overall Canadian temperatures were also above average by 1.5°C, the third-warmest year of the past half-century. The 1990s were the warmest decade on record in Canada, followed by the 1980s. Canada's warmest year of record was 1998.

Arctic Network reports that weather records from Siberia, Alaska, and northwestern Canada indicate a rise in mean temperature of approximately 1°C each decade for the last thirty years. The surface area of sea ice has been declining at a rate of 2.8 percent per decade for twenty years, a process that has accelerated since 1987 to a decadal rate of 4.5 percent. Many animals that are dietary staples of indigenous communities are ecologically dependent on the sea ice; smaller ice coverage means smaller habitats and smaller sustainable human harvests.

Dwane Wilkin, a reporter for the *Nunatsiaq News*, in Iqaluit, summarized the consequences of global warming for that area: "The good news is that sailing through the Northwest Passage will finally be a cinch. The bad news? Well, global warming will probably drive musk oxen, polar bears and Peary caribou into extinction. And other species—including humans—will face declining food sources."

(Wilkin 1997). Warmer average temperatures north of 60 degrees north latitude may mean that the Arctic Ocean will become ice-free much of the year, imperiling populations of walrus and seal that feed on creatures for whom ice means food and life.

"Our ice is getting thinner and going out. This year, we had very thin ice and it went out real fast," said Inuit elder Alexander Akeya. "It's getting more windy, and we have to hunt farther out than we used to before" (Braem 1997). Elder Jimmy Toolie had a similar observation: "Today it has changed, the ice moves all the time, and goes away quicker. We used to have it to mid-June. This year it went out in mid-May. Even in the whaling time, it seemed it went away so quick," Toolie said (Braem 1997). Another Inuit, Joe Noongwook, agreed: "This is true. We used to have huge icebergs come first before the young ice formed. We used to have thick ice from here to the [Russian] mainland to the north, but now there is thin ice. There is no hardpack ice" (Braem 1997). Several other Inuit said that the ice has been thinning especially during the last twenty years.

"Those of us who are dependent on Arctic resources know global warming is occurring," said Caleb Pungowiyi, a Yup'ik Inuit elder. "It is affecting us, and the impacts in some cases are quite severe" ("Global Warming" 1998). Pungowiyi lives in Savoonga, a village of about 500 people, near Nome, Alaska. The people of Savoonga make their livings mainly from the sea, harvesting seals, walrus, polar bear, whales, salmon, trout, whitefish, and other marine species.

Yup'ik elders in the small coastal Alaskan village of Kipnuk believe the village is sinking because of warming permafrost. Buildings in the village show signs of an unstable ground surface, signs that are consistent with those of an area that sits above thawing land. In Kotzebue, Alaska, the hospital has been relocated because it was sinking into the ground.

Rising seas and coastal erosion directly threaten Tuktoyaktuk, a Dene and Inuit community at the edge of the Arctic Ocean. Ice that once protected the coast has receded out to sea. Extensive erosion washed away the school and has forced the village to relocate many other structures.

The bay on which Iqaluit sits is named for Martin Frobisher, an early explorer who sought a northwest passage from Europe to Asia. With the polar ice cap in the Canadian Arctic melting, the fabled Northwest Passage is now opening. Commercial shippers are eyeing the prospect of cutting more than four thousand miles off their usual trip through the Panama Canal, bringing one more finger of the industrial south to the Arctic.

In the Canadian Inuit town of Inuvik, ninety miles south of the Arctic Ocean, the temperature rose to 91°F on June 18, 1999, a type of weather unknown to living memory in the area. "We were down to our T-shirts and hoping for a breeze," said Richard Binder, fifty, a local whaler and hunter. Along the Mackenzie River, according to Binder, "Hillsides have moved even though you've got trees on them. The thaw is going deeper because of the higher temperatures and longer periods of exposure." In some places near Binder's village, the thawing earth has exposed ancestral graves, and remains have needed to be reburied (Sudetic 1999, 106). Some hunters say that seals have moved farther north, killer whales are eating sea otters, and beaver are proliferating, none of which would happen if rivers and ponds were freezing to usual depths.

## LAND OF MELTING ICE AND BURNING TUNDRA

As tundra and permafrost thaws, it releases stored carbon dioxide and methane, which further raises the atmosphere's level of greenhouse gases. During Alaska's unusually warm, dry winter of 2000–2001, the snowless tundra caught fire along Norton Sound, a possible precursor of larger smoldering fires that could further accelerate global warming. There is a palatable fear among climate scientists that such "biotic feedbacks" in the oceans and on lands could release large amounts of carbon dioxide and methane that are now stored in earth.

Further evidence that the Earth's frozen reaches already are releasing additional greenhouse gases into the air has been provided by M. L. Goulden and colleagues (1998), who studied boreal forests, finding evidence that "provides clear evidence that carbon dioxide locked into permafrost several hundred to seven thousand years ago is now being given off to the atmosphere as warming climate melts the permafrost" (Goulden et al. 1998, 214). These researchers investigated black spruce forest and found that carbon dioxide was being emitted into the atmosphere from well beneath the biologically active layer containing moss and tree roots (Goulden et al., 1998). The same is true in many cases for methane. According to Davis (2001), this "relict" carbon dioxide represents a massive source since it is estimated that the carbon dioxide contained in the seasonally and perennially frozen soils of boreal forests is 200 to 500 billion metric tons, enough if all released to increase the atmosphere's concentration of carbon dioxide by 50 percent. Hence, it is possible that the release of carbon dioxide from melting permafrost during warming, or locking it into newly frozen soil during cooling, may accelerate climate change. (Davis 2001, 270)

Coastal hunters above the Arctic Circle in Alaska say they are definitely seeing a trend: The ice regularly comes a month later than it did twenty years ago, and roughly two months later than thirty years ago. Ice also breaks up earlier than previously, and so hunting seasons are becoming shorter.

"Sea ice is a pretty sensitive indicator," said Gunter Weller, a professor of geophysics at the University of Alaska at Fairbanks. "It doesn't take much [temperature change] to make a change in the ice" (Kizzia 1998). Weller noted that researchers on the National Science Foundation's ice ship *Sheba* found the polar ice cap north of Alaska to be considerably thinner than previous studies had indicated. They couldn't find an ice floe thick enough to anchor their icebreaker safely. Lack of sea ice causes a feedback warming effect because open ocean absorbs more of the sun's energy than ice and snow.

The environmental group Greenpeace and the Arctic Network, a nonprofit conservation group in Alaska, have been interviewing indigenous people of the Arctic about the condition of the ice with which they live every day. These anecdotal accounts support statistics indicating that Alaska's climate is warming more rapidly than any other place on Earth from an increase in carbon dioxide and other greenhouse gases in the atmosphere.

Hudson Bay could be ice-free by the middle of this century. "We're starting to warm up very, very fast," said Peter Scott, scientific coordinator of the Churchill [Manitoba] Northern Studies Center ("Hudson Bay" 2001). Scott said that climate change is bringing some species to the Arctic that have never been seen there before. "We may get southern animals. We're seeing a lot of moose now. . . . We now have moose in Churchill, where 20 years ago moose virtually didn't exist" ("Time to Act" 2001).

Steven Kooneeliusie and the other Inuit who hunt caribou, seal, and other animals say the signs of a gradual increase in temperatures are everywhere around them. "When I went hunting years ago,

I used to wear a full-length caribou skin coat, but now I just wear a light parka. It is so hot these days my snowmobile often overheats," Kooneeliusie said in the small town of Pangnirtung, about a hundred miles north of Iqaluit, astride the Arctic Circle (Ljunggren 2000).

Sustainable development minister Peter Kilabuk, based in Iqaluit, grew up in Pangnirtung. "I know when I was probably eight or ten [years of age] the ice wouldn't go out until July, sometimes not until the second week of July. But over the last few years we've seen the ice go out as early as May," he said. (Some years, such as 2001, the ice melted, as in previous years, in late June). "To us, the effects are real. Climate change is here and it's a real cause for concern" (Ljunggren 2000). Grizzly bears and wolverines have been moving north, said François Rainville at Environment Canada's office in Iqaluit. "There are insects and birds which have not been seen here before. There is an impact. People are seeing change," he said. Last year one Iqaluit woman reported seeing a robin (Ljunggren 2000).

Pangnirtung is the main gateway to Auyuittuq National Park— "the land that never melts," the most northerly national park in North America. Established in 1972, the park covers 19,500 square kilometers of deep mountain valleys, dramatic fjords, ancient glaciers, and spiny peaks. Auyuittuq (pronounced "ow-you-ee-tuk") is famed for its enormous glaciers. Local people say that even these glaciers are slowly melting. "The glaciers have receded over the last 10 years, and the ice is much worse," hunter Solomon Nakoolak said (Ljunggren 2000).

"If you look at weather patterns for the next 40 years, you'll see a significant warming in the Eastern Arctic, but mainly in the winter," said Brian Paruk, a meteorologist with Environment Canada's Arctic Weather Centre (George 2000b). He said the two main influences on Arctic weather to date have been the "incredible retreat" of the sea ice and the resulting increase in precipitation. More open water produces more rain. As a result of the lower temperature of the open water, summer temperatures have declined in many coastal areas (particularly along the Davis Strait) as they have risen inland.

## HUNGRY POLAR BEARS AND A DYING SEA-ICE ECOLOGY

The polar bear is one of the most visible casualties of a deteriorating ice ecology in the Arctic. Polar bears find their food supply shrinking with the melting sea ice even as their bodies acquire higher levels of organochlorines with each generation. In the pre–global warming days, polar bears had their own food sources and usually went about their business without trying to swipe food from humans. Beset by the melting of the ice that once supplied them with protein, hungry polar bears are coming into contact with people more frequently. In Churchill, Manitoba, bears waking from their winter's slumber have found Hudson's Bay ice melted. Instead of making their way onto the ice in search of food, the bears walk along the coast until they get to Churchill, where they get in the way of motor traffic and pillage the town dump. Churchill now has a holding tank for wayward polar bears that is larger than its jail for people (Krajick 2001; Linden 2000). Around the Arctic, polar bears have been observed anxiously pacing the shorelines, waiting for the food-bearing ice that arrives later and later.

Canadian Wildlife Service scientists reported that during December 1998 polar bears around Hudson Bay were 90 to 220 pounds lighter than thirty years ago, apparently because earlier ice melting has given them less time to feed on seal pups. When sea ice fails to reach a particular area, the entire ecological cycle is disrupted. When the ice melts, the polar bears can no longer use it to hunt for ring seals, many of which also have died, having had no ice on which to haul out.

During 1999 a hungry polar bear mother and a cub surprised some picnickers just outside Pangnirtung. The bears were shot after they could not be scared away. On another occasion, a school group from Arctic Bay, on the Arctic Ocean at the northern tip of Baffin Island, was camped outside town. The children were playing radios and otherwise making too much noise to notice the polar bear that entered their camp and stuck its head into the tent of the biology teacher. Awakened by the uproar, the group's guide, Simon Qamanariq, stumbled from his tent to find the bear between him and his *qamatiik*, the sledge pulled by his snowmobile where he kept his rifle. All there was at hand was a tea kettle, which Simon threw with a clatter at the polar bear's feet. While the bear was distracted, Simon ran for his gun and got off a few rounds, sending the bear running (Rauber n. d.).

## AN ALASKAN VILLAGE ERODES INTO THE ARCTIC OCEAN

Six hundred people in the Inuit village of Shishmaref, on the Chukchi Sea, about sixty miles north of Nome, have been watching their village erode into the sea. The permafrost that had reinforced its coast is thawing. "We stand on the island's edge and see the remains of houses fallen into the sea," wrote Anton Antonowicz of the *London Daily Mirror.* "They are the homes of poor people. Half-torn rooms with few luxuries. A few photographs, some abandoned cooking pots. Some battered suitcases" (Antonowicz 2000, 8).

Percy Nayokpuk, a village elder, runs the local store, which now perches dangerously close to the edge of the advancing sea. "When I was a teenager, the beach stretched at least 50 yards further out," said Percy. "As each year passes, the sea's approach seems faster (Antonowicz 2000, 8). Five houses have washed into the sea; the U.S. Army has moved or jacked up others. The villagers have been told they will soon have to move.

Year by year, the hunting season, which depends on the arrival of the ice, starts later and ends earlier. "Instead of dog mushing, we have dog slushing," said Clifford Weyiouanna, 58, a reindeer herder near Shishmaref (Antonowicz 2000, 8). Villagers have been catching fish, such as flounder, which are usually associated with warmer water.

By the summer of 2001, the encroaching sea was threatening a rusty fuel-tank facility holding 80,000 gallons of gasoline and stove oil. "Several years ago," observed Kim Murphy of the *Los Angeles Times,* "The tanks were more than 300 feet from the edge of a sea-side bluff. But years of retreating sea ice have sent storm waters pounding, and today just 35 feet of fine sandy bluff stands between the tanks and disaster" (Murphy 2001, A1). Seawater is lapping near the town's airport runway, its only long-distance connection to the outside. At last count, five houses had been washed into the sea. Several more are threatened. The town's drinking-water supply also has been inundated by the sea. The sea is eight feet from cutting the town's main road and threatens to wash the town dump out to sea.

Shishmaref hunters are being forced to search up to two hundred miles from town for walrus because of retreating ice. They also now use boats to hunt seals that they used to track over ice. "This year the ice was thinner, and most of the year at least part of the ice was

open. We don't normally see open water in December," said Edwin
Weyiouanna, an artist who has lived most of his life on the Chukchi
Sea (Murphy 2001, A1). In earlier years, the sea was more likely to
be frozen during much of the stormy winter season. With warming,
the erosive wind-whipped ocean corrodes Shishmaref's waterfront.
The town's residents have come to fear the full moon, with its unusually high tides.

Gunter Weller, director of the Center for Global Change and Arctic
System Research at the University of Alaska in Fairbanks, said
mean temperatures in the state have increased by 5°F in the summer and 10°F in the winter over the last thirty years. Moreover, the
Arctic ice field has shrunk by 40 percent to 50 percent over the last
few decades and has lost 10 percent of its thickness, studies show.
"These are pretty large signals, and they've had an effect on the
entire physical environment," Weller said (Murphy 2001, A1).

Increasing coastal erosion isn't limited to Shishmaref; it is now
general on the shores of the Bering Sea, as increasing storm surges
crush ice packs that retreat several weeks earlier than thirty years
ago. Sea-ice extent in the Arctic has decreased Arctic-wide by 0.35
percent per year since 1979. During the summer of 1998, record
reduction of sea-ice coverage was observed in the Beaufort and
Chukchi seas (Maslanik, Serreze, and Agnew 1999).

In Tuktoyaktuk, a town near the mouth of the Mackenzie River,
several buildings have been lost to erosion by the sea. In addition
to a reduction in Bering Sea ice cover, more precipitation is falling
in many areas of Alaska. During the fall of 1998, sea ice formed in
northern Alaska more than a month later than usual, postponing
the annual seal and walrus hunt.

Average temperatures at the mouth of the Mackenzie River were
9°F above long-term averages during 1998. Several decades ago,
miners in the area deposited toxic wastes in ponds that were expected to remain frozen (and the toxic materials sealed by the permafrost). With warmer temperatures, some of these toxic dumps
may thaw and leak.

In Alaska, where roughly 80 percent of glaciers are receding, forests of dead spruce surround Anchorage, a casualty of a spruce-
beetle epidemic caused at least in part by rising temperatures,
which accelerate the insect's life cycle. Large patches of Alaskan
forests have been described as drowning and turning gray as thawing ground sinks under them. Trees and roadside utility poles,

destabilized by thawing, lean at crazy angles. The warming has contributed a new phrase to the English language in Alaska—"the drunken forest."

The Sierra Club of Canada created a television commercial that highlights the impact of global warming on the Arctic, including the disastrous effect of receding sea ice on polar bear populations and the hardship that increasing snowfall brings to the endangered Peary caribou. "Our way of life is on the edge of extinction. Plants and animals are dying," said Rosemarie Kuptana, a former president of the Inuit Circumpolar Conference, who was recruited to host the thirty-second spot (George 2000b).

On Earth Day 2000 Inuit brought their accounts of dramatic warming to urban audiences in southeastern Canada. Kuptana told a press conference in Ottawa that experienced hunters have fallen through unusually thin ice. Three men had recently died this way. Never-before-seen species (including mosquitoes, robins, barn swallows, beetles, and sand flies) have appeared on Banks Island, in the Arctic Ocean about eight hundred miles northwest of Fairbanks, Alaska. Growing numbers of Inuit are suffering allergies from white-pine pollen that recently, for the first time, reached Sachs Harbour, on western Banks Island, along the Beaufort Sea.

"If this rate of change continues, our lifestyle may forever change, because our communities are sinking with melting permafrost and our food sources are . . . more difficult to hunt," said Kuptana (Duffy 2000, A13). Sachs Harbour itself is slowly sinking during the summer into a muddy mass of thawing permafrost. Born in an igloo, Kuptana has been an Inuit weather watcher for much of her life (she was forty-six years of age in 2000). Her job was to scan the morning clouds and test the wind's direction to help the hunters decide whether to go out and what everyone should wear. Now she gathers observations for international weather-monitoring organizations. "What happens in the Arctic environment is what is in store for other regions of the world," said Kuptana (Duffy 2000).

"We can't read the weather like we used to," said Kuptana (Duffy 2000), who grew up in Sachs Harbour. She said that autumn freezes now occur a month later than they used to; spring thaws come earlier as well. Residents of Sachs Harbour still suffer through winters that most people from lower latitudes would find chilling, with temperatures of minus 40°F or lower. While such temperatures were once commonplace, however, they now are rare.

"The permafrost is melting at an alarming rate," said Kuptana (Herbert 2000). Foundations of homes in Sachs Harbour have been cracking and shifting. Kuptana expressed fear that the community will soon slide into the Beaufort Sea because once-solid mud is now thawing earlier and freezing later.

"What's scary is the uncertainty," she said. "We don't know when to travel on the ice and our food sources are getting further and further away" (Herbert 2000). Tracking animals on receding ice and mushy ground poses new problems for Inuit hunters, some of whom have been reporting sunburns for the first time. "We have no other sources of food, the people in my community are completely dependent on hunting, trapping and fishing," said Kuptana. "We have no means of adapting to a different environmental reality, and that is why our situation is so critical" (Herbert 2000).

At Sachs Harbour, sea ice is thinner and now drifts far away during the summer, taking with it the seals and polar bears upon which the village's Inuit residents rely for food. Many young seals are starving to death because prematurely melting sea ice separates them from their mothers. In the winter the sea ice often is thin and broken, making travel dangerous for even the most experienced hunters. In the fall, storms have become more frequent and often more severe, making boating difficult. Thunder and lightning have been seen for the first time, arriving with another type of weather that is new to the area—dousing summer rainstorms.

"When I was a child, I never heard thunder or saw lightning, but in the last few years we've had thunder and lightning," Kuptana said. "The animals really don't know what to do because they've never experienced this kind of phenomenon" ("Thunder Storms" 2000). "We don't know when to travel on the ice and our food sources are getting further and further away," says Kuptana. "Our way of life is being permanently altered" (Knight 2000). Andy Carpenter of Sachs Harbour observed that "Different kinds of birds are coming up. There's other species of fish that we've never gotten before, like the pink salmon" ("Scientists Look" 2001).

During 1997 and 1998, a team of 160 scientists lived on the ice of the Arctic Ocean, about 500 kilometers from the North Pole. Their headquarters aboard the Canadian icebreaker *Des Groseilliers* was converted into a floating laboratory. Biologist Harold Welch, a researcher on the team, said a comparison with studies done thirty years ago documents significant melting of the polar ice cap.

"With climate warming, we get a thinning of the arctic ice pack and more open water, and a longer summer with less snow cover, which allows more light to come in," Welch said (Bourrie 1998). Scientists reported that the ice pack has become so thin that the research camp nearly was lost several times as the ice around it split apart. "A couple of times, we saw haze in the atmosphere and the sun appearing as a red ball because of forest fires and other fires in Asia. But much more important, we see contaminants, pollution, that comes from Eurasia, primarily," Welch said (Bourrie 1998).

Greenpeace activists during 1997 gathered information about global warming from Yup'ik and Inupiat Inuit. They embarked on the organization's 150-foot ship *Arctic Sunrise,* on a route north to Barrow along Alaska's Arctic coast. The activists reported that Savoonga whalers brought home no whales in 1997 because of a quick and early breakup of ice that forced them to come home earlier than usual. Inuit said that for several years they have been routinely catching fish heretofore not considered native to the Arctic, including chum and pink and silver salmon. Like observers across the Arctic, Inuit interviewed by Greenpeace said they had observed birds that are new to higher latitudes, such as various types of swallows, finches, and robins.

A Canadian government report, *Responding to Global Climate Change in Canada's Arctic,* released late in 1997, did its best to put a positive spin on global warming in the Arctic. Reading closely, however, one realizes that most of the benefits accrue to industrial society's incursion into the Arctic. The report noted that reduced ice cover would aid offshore oil and gas production and extend the shipping season. In the western Arctic, said the same report, "New agricultural opportunities would arise" (Wilkin 1997).

Most of global warming's debits are accruing to the Inuit, the animals of the region, and the traditional relationship of the two. The report expects that precipitation will increase as much as 25 percent with the advent of rapidly rising temperatures during a century. When the precipitation falls as snow and ice, caribou and other grazing animals will have a harder time finding food during lean winter months. The study anticipates that heavier snow cover will lead to smaller, thinner animals that will be forced to go further afield for food in winter. The same animals will be plagued by increasing numbers of insects during warmer summers. Environmental stresses during the next century could lead to "complete reproductive failure of the caribou in a worst-case scenario" (George 2000a).

The Canadian study forecasts that winter temperatures in the mainland of the Canadian Arctic will rise by 5 to 7°C within a century, unless greenhouse-gas emissions are drastically curtailed worldwide. Such temperatures would melt more than half the existing permafrost in the Canadian Arctic, according to the same report. Some fear that as the ground heats up airport runways and buildings could crack (Wilkin 1997).

Other impacts of global warming include "wildlife habitat destruction, melting permafrost which destabilizes northern homes, roads and pipelines, large-scale coastal erosion, and insect outbreaks" (Ralston 1996). The report suggests that an increase in the incidence of forest fires in the area may be attributed to rising temperatures (Ralston 1996). Forest fires are adding even more greenhouse gases to the atmosphere.

## THE ACCUMULATION OF MERCURY IN THE ARCTIC

In addition to accumulation of persistent organic pollutants, springtime also brings mercury-laced precipitation to the polar regions. The mercury arrives in much the same manner as dioxins and PCBs, via prevailing currents in the ocean and atmosphere. Fossil-fuel combustion releases about 6,500 tons of mercury vapor into the atmosphere every year ("Mercury" 2002). "In just five months during the spring each year," according to a report by the Environment News Service (ENS), "The northernmost coast of Alaska, a region that includes the Arctic National Wildlife Refuge, receives more than double the amount of mercury that falls on the northeastern United States in a year" ("Mercury" 2002). Similar pollution also occurs each year in the Antarctic.

Two studies in the March 15, 2002, edition of *Environmental Science and Technology*, a journal published by the American Chemical Society, presented new information about what some scientists had begun to call the "mercury sunrise," which was first reported in the Arctic during 1998 (Ebinghaus et al. 2002; Lindberg et al. 2002). The first sunrise of spring initiates a series of chemical reactions that dump mercury out of the atmosphere into polar snow pack, from which it is released into the food web during the brief polar growing season.

Ralf Ebinghaus, an environmental chemist at the Institute for Coastal Research in Geesthacht, Germany, gave a description of the mercury sunrise at Germany's Neumayer research station in Antarctica based on measurements between January 2000 and January 2001. According to an ENS report, Steve Lindberg, a research corporate fellow at the Oak Ridge National Laboratory in Tennessee, "presented the first evidence of how this phenomenon leads to elevated mercury accumulation in snow, as measured at the National Oceanic and Atmospheric Administration's Climate Monitoring and Diagnostic Laboratory in Barrow, Alaska" ("Mercury" 2002).

A research team led by Lindberg wrote that

Our data from Barrow, Alaska, at 71 degrees north, show that rapid, photochemically driven oxidation of boundary-layer HgO [gaseous elemental mercury] after polar sunrise, probably by reactive halogens, creates a rapidly depositing species of oxidized gaseous mercury in the remote Arctic troposphere. . . . This mercury accumulates in the snow pack during polar spring at an accelerated rate in a form that is bio-available to bacteria and is released with snowmelt during the summer emergence of the Arctic ecosystem. Evidence suggests that this is a recent phenomenon that may be occurring throughout the earth's polar regions." (Lindberg et al., 2002, 1245)

The ENS report described findings of the Lindberg team, which may help explain why mercury levels in Arctic seabirds, seals, and beluga whales have increased over the last twenty years, even as global atmospheric emissions of mercury have declined ("Mercury" 2002). The mercury sunrise is said by the scientists to depend on ultraviolet light, open water, and active sea ice, all of which have been increasing in polar regions during the last few decades. According to ENS, the Barrow data indicates that the polar sunrise triggers a series of chemical reactions that convert elemental mercury vapor into a soluble form of oxidized gaseous mercury that accumulates in the polar snow pack ("Mercury" 2002). Ebinghaus and Lindberg estimated that between 50 and 300 tons of mercury were being dumped from the atmosphere into these polar environments each year by the year 2001.

## STRATOSPHERIC OZONE DEPLETION IN THE ARCTIC

Along with toxic pollutants and accelerating global warming, the Inuit (and the Arctic ecosystem generally) are facing another im-

ported threat from the industrial south. Stratospheric ozone depletion in northern latitudes and resulting increases in ultraviolet (UV)
radiation have increased markedly during the past decade, with
some sectors of the Arctic experiencing a 40 to 60 percent increase
in UV radiation (Taalas et al. 1996, 1997).

The National Science Foundation maintains a polar UV monitoring Network, which has observed

In the last few decades, particularly in the 1990s, anthropogenic influences
on the natural ozone layer have resulted in severe ozone depletion in polar
regions. . . . During March 1997, stratospheric ozone in the Arctic reached
all-time lows with losses due to halogen-related chemistry rivaling the Antarctic losses observed since the infamous "ozone hole" was discovered in
1985. Three-dimensional, coupled physical-chemical models indicate that
depletion of stratospheric ozone in the Arctic will increase in the future as
climate change cools the stratosphere and stabilizes the polar vortex during
the period when stratospheric chlorine and bromine from human-made hal-
ogen compounds will remain high for the next 20 to 30 years. (Bodhaine,
Dutton, and Tatusko 2001)

The same report continued "It is during the spring, when ozone depletion is greatest, that biological systems are most susceptible to
UV damage: natural protective measures, such as pigmentation and
thickening of leaves, have not had time to develop, and fish larvae
are most exposed" (Bodhaine et al., 2001).

During the 1990s, the ozone column over the Antarctic failed to
heal as projected after the production of ozone-destroying chlorine
compounds was banned under the Montreal Protocol. According to
scientific analysis developing early in the twenty-first century, the
warming of the near-surface atmosphere (the lower troposphere) is
related to the cooling of the stratosphere, which drives ozone depletion at that level. Increases in carbon dioxide and other greenhouse
gases near the Earth's surface acts as a blanket, trapping the heat.
By 1998 the Antarctic ozone hole reached a new record size roughly
the size of the continental United States. A year later, it was roughly
the size of North America. The area of severely depleted ozone stabilized during the next few years.

Steve Hipskind, atmospheric and chemistry dynamics branch
chief at NASA's Ames Research Center, Moffett Field, California, has
been quoted as saying that chlorine atoms must use clouds as a
platform to destroy stratospheric ozone ("Arctic Region" 2000, 4).
Clouds form more frequently in the stratosphere at low tempera-

tures, most notably below minus 107°F. Ice crystals, which form as part of polar stratospheric clouds, assist the chemical process by which ozone is destroyed. During the winter of 1999–2000, temperatures in the stratosphere over the Arctic were recorded as low as minus 118°F (the lowest on record), forming the necessary clouds to allow accelerated ozone depletion.

As temperatures fall in the Arctic stratosphere, the ozone column has been depleted there as well. During April 2000 scientists from the United States and Europe said that more than 60 percent of the Arctic ozone layer about eleven miles above the Earth had vanished during the winter because of record stratospheric cold and continued pollution—one of the most substantial ozone losses ever recorded there. Evidence is accumulating that indicates that ozone depletion will continue to be a problem in the Arctic until the industrialized world substantially reduces its consumption of fossil fuels. Even then, it will take about thirty to fifty years for various feedback loops to reverse the rapid rise in near-surface temperatures. Only after that will the thermal imbalance between the lower and upper atmosphere correct itself, allowing the ozone shield to restore itself. In the meantime, the survival of the Inuit, the polar bears, and the rest of the Arctic ecosystem hang in the balance.

## REFERENCES

Antonowicz, Anton. 2000. "Baking Alaska: As World Leaders Bicker, Global Warming Is Killing a Way of Life." *London Daily Mirror,* November 28.
"Arctic Region Quickly Losing Ozone Layer." 2000. *Omaha World-Herald,* April 6.
Bodhaine, Barry, Ellsworth Dutton, and Renee Tatusko. 2001. "Assessment of Ultraviolet (UV) Variability in the Alaskan Arctic." Cooperative Institute for Arctic Research, University of Alaska and NOAA. March 6. http://www.cifar.uaf.edu/ari00/bodhaine.html.
Bourrie, Mark. 1998. "Global Warming Endangers Arctic." Interpress Service, October 14. http://www.muskox.com/news/news1.html.
Braem, Nicole M. 1997. "Greenpeace Activists Visit Yup'ik and Inupiat Villages in Alaska to Gather Information about Global Warming." *Nunatsiaq News,* August 8. http://www.nunanet.com/~nunat/week/70808.html.
Brooke, James. 2000. "Canada's Bleak North Is Fertile Ground for Suicide." *Canadian Aboriginal News,* December 18. http://www.canadian aboriginal.com/health/health21a.htm.

Cadbury, Deborah. 1997. *Altering Eden: The Feminization of Nature.* New York: St. Martin's Press.

Calamai, Peter. 2000. "Chemical Fallout Hurts Inuit Babies." *Toronto Star,* March 22. http://irptc.unep.ch/pops/newlayout/press_items.htm.

Cone, Marla. 1996. "Human Immune Systems May Be Pollution Victims." *Los Angeles Times,* May 13.

"Conservation of Arctic Flora and Fauna: Arctic Climate Impact Assessment. An Assessment of Consequences of Climate Variability and Change and the Effects of Increased UV in the Arctic Region: A Draft Implementation Plan." 1999. October 22. http://www.grida.no/caff/acia.htm.

Davis, Neil. 2001. *Permafrost: A Guide to Frozen Ground in Transition.* Fairbanks: University of Alaska Press.

Dewailly, E., P. Ayotte, S. Bruneau, S. Gingras, M. Belles-Isles, and R. Roy. 2000. "Susceptibility to Infections and Immune Status in Inuit Infants Exposed to Organochlorines." *Environmental Health Perspectives* 108: 205–11.

Dewailly E., S. Bruneau, C. Laliberte, M. Belles-Iles., J. P. Weber, and R. Roy. 1993. "Breast Milk Contamination by PCB and PCDD/Fs in Arctic Quebec. Preliminary Results on the Immune Status of Inuit Infants." *Organohalogen Compounds* 13: 403–6.

Dewailly E., S. Dodin, R. Verreault, P. Ayotte, L. Sauve, and J. Morin. 1993. "High Organochlorine Body Burden in Breast Cancer Women with Oestrogen Receptors." *Organohalogen Compounds* 13: 385–88.

Dewailly E., J. J. Ryan, C. Laliberte, S. Bruneau, J. P. Weber, S. Gingras, and G. Carrier. 1994. "Exposure of Remote Maritime Populations to Coplanar PCBs." *Environmental Health Perspectives* 102, suppl. 1: 205–9.

Duffy, Andrew. 2000. "Global Warming Causing Arctic Town to Sink, Says Inuit Leader 'Warning Signal.'" *Montreal Gazette,* April 18. http://www.climateark.org/articles/2000/2nd/glwatosi.htm.

Ebinghaus, Ralf, Hans H. Kock, Christian Temme, Jürgen W. Einax, Astrid G. Löwe, Andreas Richter, John P. Burrows, and William H. Schroeder. 2002. "Antarctic Springtime Depletion of Atmospheric Mercury." *Environmental Science and Technology* 36, no. 6 (March 15): 1238–44.

George, Jane. 2000a. "Global Warming, Inbreeding Threaten Nunavut Muskoxen." *Nunatsiaq News,* April 14. http://www.nunatsiaq.com/archives/nunavut000430/nvt20414_08.html.

———. 2000b. "Sierra Club Focuses on Arctic Global Warming." *Nunatsiaq News,* January 31. http://www.nunatsiaq.com/archives/nunavut000131/nvt20121_04.html.

"Global Warming Worries Native Americans." 1998. *Deseret News,* November 27. http://www.desnews.com/cit/071en14f.htm.

Goulden, M. L., S. C. Wofsy, J. W. Harden, S. E. Trumbone, P. M. Crill, S. T. Gower, T. Fries, B. C. Daube, S.-M. Fan, D. J. Sutton, A. Bazzaz, and J. W. Munger. 1998. "Sensitivity of Boreal Forest Carbon Dioxide to Soil Thaw." *Science* 279 (January 9): 214–17.

Herbert, H. Josef. 2000. "Inuit Say They Are Witness to Global Warming in the Arctic." *Milwaukee Journal-Sentinel,* November 19. http://www.jsonline.com/alive/news/nov00/warming20111900.asp?format=print.

Hill, Miriam. 2001. "Iqaluit's Waste Woes Won't Go Away; City Sets Up Bins Where Residents Can Dump Plastics, Metal." *Nunatsiaq News,* July 27.

"Hudson Bay Ice-Free by 2050, Scientists Say." 2001. Canadian Broadcasting Corporation. March 14. http://north.cbc.ca/cgi-bin/templates/view.cgi?/news/2001/03/14/14hudsonice.

"Iqaluit Drunks Keep Cops' Hands Full." 2001. Canadian Broadcasting Corporation North/News. March 16. http://north.cbc.ca/cgi-bin/templates/view.cgi?/news/2001/03/16/16iqalcohol.

Jacobson, Joseph L., and Sandra W. Jacobson. 1996. "Intellectual Impairment in Children Exposed to Polychlorinated Biphenyls *in Utero.*" *New England Journal of Medicine* 335, no. 11: 783–89.

Johansen, Bruce E. 2000. "Pristine No More: The Arctic, Where Mother's Milk Is Toxic." *Progressive,* Dec., 27–29.

Kizzia, Tom. 1998. "Seal Hunters Await Late Ice." *Anchorage Daily News,* November 28. http://www.adn.com/stories/T98112872.html.

Knight, Danielle. 2000. "Inuit Tell Negotiators of Climate Change Impact." Interpress Service *World News,* November 16. http://www.oneworld.org/ips2/nov00/01_44_005.html.

Krajick, Kevin. 2001. "Arctic Life, on Thin Ice." *Science* 291 (January 19): 424–25.

Lamb, David Michael. n.d. "Toxins in a Fragile Frontier." Transcript, Canadian Broadcasting Corporation News. http://cac.ca/news/indepth/north.

Lindberg, Steve E., Steve Brooks, C.-J. Lin, Karen J. Scott, Matthew S. Landis, Robert K. Stevens, Mike Goodsite, and Andreas Richter. 2002. "Dynamic Oxidation of Gaseous Mercury in the Arctic Troposphere at Polar Sunrise." *Environmental Science and Technology* 36, no. 6 (March 15): 1245–56.

Linden, Eugene. 2000. "The Big Meltdown: As the Temperature Rises in the Arctic, It Sends a Chill around the Planet." *Time,* September 4, 52.

Ljunggren, David. 2000. "Effects of Global Warming Clear in Canada Arctic." Environmental News Network, April 20. http://www.enn.com/enn-subsciber-news-archive/2000/04/04202000/reu_arctwarm_12170.asp.

Maslanik, J. A., M. C. Serreze, and T. Agnew. 1999. "On the Record Reduction in 1998 Western Arctic Sea Ice Cover." *Geophysical Research Letters* 26, no. 13: 1905–8.

"Mercury Contaminates Polar Regions." 2002. Environment News Service, March 21. http://ens-news.com/ens/mar2002/2002L-03-21-09.html #anchor1.

Mofina, Rick. 2000. "Study Pinpoints Dioxin Origins: Cancer-Causing Agents in Arctic Aboriginals' Breast Milk Comes from U.S. and Quebec." *Montreal Gazette,* October 4.

Murphy, Kim. 2001. "Front-Row Exposure to Global Warming; Climate: Engineers Say Alaskan Village Could Be Lost As Sea Encroaches." *Los Angeles Times,* July 8.

PCB Working Group, IPEN. n.d. "Communities Respond to PCB Contamination." http://www.ipen.org/circumpolar2.html.

"PD 2000 Projects." 2001. Arctic Monitoring and Research—Project directory, April 11. http://amap.no/pd2000.htm.

Pugliese, David. 2001. "An Expensive Farewell to Arms: The U.S. Has Abandoned 51 Military Sites in Canada. Many Are Polluted, and Taxpayers Are Paying Most of the $720 Million Cleanup Cost." *Montreal Gazette,* April 28.

Ralston, Greg. 1996. "Study Admits Arctic Danger." *Yukon News,* November 15. http://yukonweb.com/community/yukon-news/1996/nov15.htmld/#study.

Rauber, Paul. n.d. "On Top of the World: Midsummer on Baffin Island: Where the Sun Never Sets on the Polar Bear's Empire and Staying Alive Is a Round-the-clock Job." Sierra Club Web site. http://www.sierraclub.org/sierra/199803/baffin.html.

"Scientists Look to 'Traditional Knowledge' to Help Understand Climate Change." 2001. Canadian Broadcasting Corporation News, March 22. http://cbc.ca/cgi-bin/view?/news/2001/03/21/climate010321.

Sudetic, Chuck. 1999. "As the World Burns." *Rolling Stone,* September 2, 97–106, 129.

Suzuki, David. 2000. "Science Matters: POP Agreement Needed to Eliminate Toxic Chemicals." December 6. http://www.davidsuzuki.org/Dr_David_Suzuki/Article_Archives/weekly12060002.asp.

Taalas, P., J. Damski, E. Kyrö, M. Ginzburg, and G. Talamoni. 1997. "The Effect of Stratospheric Ozone Variations on UV Radiation and on Tropospheric Ozone at High Latitudes." *Journal of Geophysical Research* 102: 1533–43.

Taalas, P., E. Kyrö, K. Jokela, T. Koskela, J. Damski, M. Rummukainen, K. Leszczynski, and A. Supperi. 1996. "Stratospheric Ozone Depletion and Its Impact on UV Radiation and on Human Health." *Geophysica* 32: 127–65.

"Thunder Storms Are Latest Evidence of Climate Change." 2000. Associated Press Canada, November 15. http://abcnews.go.com/sections/science /DailyNews/arctic_thunder001115.html.

"Time to Act on Nunavut Climate Change." 2001. Canadian Broadcasting Corporation, March 16. http://north.cbc.ca/cgi-bin/templates/view. cgi?/news/2001/03/16/16nunmoose.

Wilkin, Dwane. 1997. "Global Warming Poses Big Threats to Canada's Arctic." *Nunatsiaq News*, November 21. http://www.nunatsiaq.com/ archives/back-issues/71121.html#6.

York, Geoffrey. 2001. "Russian City Ravaging Arctic Land." *Toronto Globe and Mail*, July 25.

# 3
# CFCs, Global Warming, and Ozone Depletion

Why has stratospheric ozone depletion over the Arctic and Antarctic accelerated despite a decade-old ban on ozone-destroying chlorofluorocarbons (CFCs)? Why did the Antarctic ozone "hole" expand during the year 2000 to an area the size of North America? Why has ozone depletion also been increasing over the Arctic? Why are ozone warnings being issued in Punta Arenas, Chile, and Ushuaia, Argentina?

Part of the answer to this riddle appears to be rising levels of greenhouse gases (carbon dioxide, methane, and others) near the surface of the Earth. In an act of atmospheric irony, warming near the surface of the Earth causes the ozone-bearing stratosphere to cool significantly. The atmosphere's existing cargo of CFCs (with a lifetime of up to a century) consume more ozone the colder it gets. An increasing level of carbon dioxide near the Earth's surface "acts as a blanket," said NASA research scientist Katja Drdla. "It is trapping the heat. If the heat stays near the surface, it is not getting up to these higher levels" (Borenstein 2000).

Thus, until humanity reduces its emissions of greenhouse gases, ozone depletion will remain a problem long after production of CFCs has ceased. By the year 2001, the ozone-depleted area over Antarctica grew, at its maximum extent, to an area the size of Africa. While the polar reaches of the Earth have been suffering the most dramatic declines in ozone density, ozone levels over most of the Earth have declined roughly 15 percent since the middle 1980s. The CFC family of synthetic chemicals do more than destroy stratospheric ozone. They also act as greenhouse gases, with several thousand times the per-molecule greenhouse potential of carbon dioxide.

As levels of greenhouse gases rise, cooling of the middle and upper atmosphere is expected to continue, with attendant consequences for ozone depletion. Guy P. Brasseur and colleagues (2000) modeled the response of the middle atmosphere to a doubling of carbon dioxide levels near the surface. Their models indicate that "A cooling of about 8°Kelvin is predicted at 50 kilometers during summer. During winter, the temperature is reduced up to 14°K. at 60 km. in the polar region" (16). Increasing levels of methane, also a greenhouse gas, enhance this effect. In addition to its properties as a greenhouse gas, "methane oxidation leads to higher water and OH concentrations in the stratosphere and mesosphere, and hence to less ozone at these altitudes" (Brasseur et al. 2000, 16).

Why does anyone need to be worried if stratospheric ozone levels decline? The major human motivation for concern is ozone's role in shielding flesh and blood residing on the surface from several frequencies of ultraviolet radiation. One of these, ultraviolet B, is energetic enough to break the bonds of DNA molecules, the molecular carriers of our genetic coding. While plants and animals are generally able to repair damaged DNA, on occasion damaged DNA molecules can continue to replicate, leading to dangerous forms of basal, squamous, and melanoma skin cancers in humans.

The probability that DNA can be damaged by ultraviolet radiation varies with wavelength, shorter wavelengths being the most dangerous. "Fortunately," writes one observer, "at the wavelengths that easily damage DNA, ozone strongly absorbs ultraviolet radiation and, at the longer wavelengths where ozone absorbs weakly, DNA damage is unlikely. But given a 10-percent decrease in ozone in the atmosphere, the amount of DNA-damaging ultraviolet radiation would be expected to increase by about 22 percent" (Newman 1998).

## A BRIEF HISTORY OF CFCS

CFCs initially raised no environmental questions when they were first marketed by DuPont Chemical during the 1930s under the trade name Freon®. Freon was introduced at a time when such questions usually were not asked. At about the same time, asbestos was being proposed as a high-fashion material for clothing, and radioactive radium was being built into timepieces so that they would glow in the dark.

CFCs were first manufactured in 1930, a year after PCBs were created. They were invented by Thomas Midgely, a chemist working for the Chevron Corporation. Their main use was as a propellant in spray cans and fire extinguishers. They also came to be used widely in compressors because, like PCBs, CFCs were useful for their stability. Before their effect on the ozone shield was detected (during the 1970s) CFCs were touted as nontoxic. Also like PCBs, CFCs accumulated in the atmosphere for several decades before negative ecological effects were suspected.

Several varieties of CFCs have been synthesized; the most often used are usually referred to as "F-11" and "F-12." By the 1970s, about 800,000 tons a year of CFCs were being injected into the atmosphere (Van Emden and Peakall 1996, 32), and some scientists began wondering what had become of them. The scientists found that CFCs were working their way into the stratosphere twenty to twenty-five miles above the Earth's surface, where they were consuming the ozone shield that protects the surface from harmful ultraviolet radiation.

By the 1970s, manufacturers in the United States were producing 750 million pounds of CFCs a year and finding all sorts of uses for them, for example, as propellants in aerosol sprays; as solvents to clean silicon chips; as home, office, and automobile air conditioning refrigerants; and as blowing agents in the manufacture of polystyrene cups, egg cartons, and containers for fast food. They were amazingly useful, cheap to manufacture, nontoxic, and nonflammable.

During June 1974 chemists Sherwood Rowland and Mario Molina reported in *Nature* that CFCs were working into the stratosphere and depleting the ozone column. Their research led to a cessation of CFC manufacture by Du Pont, and a Nobel Prize in chemistry for Rowland and Molina in 1995. The work of Rowland and Molina was theoretical (Rowland and Molina 1974). Within a few years, their theories were confirmed as scientists discovered, during the early 1980s, that CFCs were, in fact, rapidly thinning the ozone layer over the Antarctic.

Stratospheric abundance of chlorine from human sources began to increase rapidly shortly after World War II; by the year 1998, the stratosphere contained roughly six times the natural background amount provided by methyl chlorine emitted by the oceans (Crutzen 2001, 5).

At this point, opponents of CFC manufacture found themselves facing a $28-billion-a-year industry. By the time their legal manufacture was banned internationally, CFCs had been used in roughly 90 million U.S. car and truck air conditioners, 100 million refrigerators, 30 million freezers, and 45 million air conditioners in homes and other buildings.

## AN OZONE "HOLE" IS DISCOVERED

During 1985 a team of scientists working with the British Antarctic Survey reported a startling decline in "column ozone values" above an observation station near Halley Bay, Antarctica (Farman, Gardiner, and Shanklin 1985). Ozone densities had been declining slowly over the Antarctic since 1977. The size of the decline in 1985 was a shocking surprise, however, because theorists had expected stratospheric ozone levels to fall relatively evenly over the entire Earth. Rowland and Molina (1974) had expected a largely uniform worldwide decline of 1 to 5 percent. The seasonal variability of the decline was another surprise, because existing theoretical models made no allowance for it. Ozone values over Antarctica declined rapidly just as the sun was rising after the winter season.

During the middle 1980s, the cause of dramatic declines in ozone density over the Antarctic was open to debate. Some scientists suspected variability in the sun's radiational output, and others suspected changes in atmospheric circulation. A growing minority began to suspect CFCs. These chemicals were not proven suspects when, in 1987, a majority of the world's national governments signed the Montreal Protocol to eliminate CFCs.

Definite proof of CFCs' role in ozone depletion arrived shortly thereafter. J. G. Anderson, W. H. Brune, and M. H. Profitt (1989) implicated the chemistry of chlorine and explained the chain of chemical reactions (later broadened to bromides as a bit player)—the "smoking gun"—that explained why ozone depletion was so sharp and why it was limited to specific geographic areas at a specific time of the year. The temperature of the stratosphere later came to be understood as a key ingredient in the mix—the colder the stratosphere, the more active the chlorine chemistry that devours ozone. By the year 2000, according to Maureen Christie (2001), ozone depletion was "significantly affecting ozone levels throughout the Southern Hemisphere" (86).

When the Montreal Protocol was signed, the best science available was indicating that without remedial action the ozone destruction rate would increase roughly 8 to 17 percent per year (a figure cal-culated for the early 1990s); by about 1992, it was projected that anthropogenic ozone destruction would rise to thirty-six times the natural rate (Crutzen 2001, 7).

Matthew Stein, writing in *When Technology Fails,* describes how chlorine atoms come to destroy stratospheric ozone:

CFCs are normally very stable, lasting 50 to 100 years before finally break-ing down. CFCs are lighter than air and slowly migrate into the upper at-mosphere, where high-energy rays from the sun blow them apart, liberating a chlorine atom into the ozone layer. Each free atom of chlorine acts as a catalyst, breaking up thousands of ozone molecules before finally reacting with something else, which removes it from circulation. (Stein 2000, 17)

In the stratosphere, CFCs release reactive chlorine and bromine that destroy ozone. The destruction is catalytic, as each chlorine or bromine radical destroys thousands of ozone molecules. Stein also pointed out that "In addition to CFCs, each new launch of America's Space Shuttle delivers tons of chlorine molecules from spent rocket fuel directly to the upper atmosphere, where they eat away at the planet's ozone shield" (Stein 2000, 17).

The decline of stratospheric ozone is striking when viewed on a graph with any sense of historical proportion. As little as a decade or two will do. Until the 1990s, in the Arctic, springtime ozone levels ranged around 500 Dobson units. By the year 2001 they were av-eraging 200 to 300; in the Antarctic, in the days before the "ozone hole" (about 1980) Dobson-unit values ranged from about 250 to 350; by the year 2000, they ranged from 100 to 200. Scientists usu-ally call an area of the stratosphere ozone-depleted when its D.U. level falls below roughly 200.

Measurements of stratospheric ozone above Arrival Heights, near Scott Base in Antarctica, reached the lowest level ever recorded, 124 Dobson units, on September 30, 2000. The Antarctic ozone "hole" formed earlier and endured longer during September and October of 2000 than ever before. Figures from NASA satellite measure-ments showed that the hole covered an area of approximately 29 million square kilometers in early September, exceeding the previ-ous record from 1998. The record size persisted for several days. Ozone levels fell below 100 D.U. for the first time in some areas. The area cold enough to materially enhance ozone depletion also grew to 10 to 20 percent more surface area than any previous year. By the year 2000 the ozone-depleted zone was coming closer to New Zealand, where usual springtime ozone levels average about 350 D.U. During spring 2000, ozone levels went as low as 260 D.U.

The atmosphere's cargo of CFC's did not dissipate immediately, of course. Infrastructure using CFCs was replaced gradually, over several years; CFCs also enter the atmosphere in a number of ways not banned by the Montreal Protocol, some of them as prosaic (and widespread) as frying food on a Teflon®-coated surface. Heated Teflon also releases into the atmosphere small amounts of fluorocarbons, which are potent greenhouse gases.

Paul J. Crutzen asserts that problems with the stratospheric ozone layer could have been much worse if chemists had developed substances based on bromine, which is 100 times as dangerous for ozone atom for atom as chlorine. "This brings up the nightmarish thought that if the chemical industry had developed organobromine compounds instead of the CFCs—or, alternatively, if chlorine chemistry had behaved more like that of bromine—then without any preparedness, we would have faced a catastrophic ozone hole everywhere and in all seasons during the 1970s, probably before atmospheric chemists had developed the necessary knowledge to identify the problem" (Crutzen 2001, 10). Given the fact that no one seemed overly worried about this problem before 1974, writes Crutzen, "We have been extremely lucky." This shows, he writes, "That we should always be on our guard for the potential consequences of the release of new products into the environment . . . for many years to come" (Crutzen 2001, 10).

## POLAR STRATOSPHERIC CLOUDS

The chlorine and bromine reactive forms are exacerbated by chemical reactions that take place most efficiently on the surfaces of cloud particles. Before the advent of fossil-fuel-forced global surface warming (and stratospheric cooling) students of atmospheric sciences were taught that clouds form in the stratosphere very rarely, if at all. In the brave new world of atmospheric chemistry in an era of whole-Earth pollution, however, polar stratospheric clouds even have their own acronym: "PSCs."

Mesospheric clouds form above the poles at altitudes of eighty-two to eighty-six kilometers. They have been increasing in brightness during the last four decades, and these changes may be linked to anthropomorphic emissions of carbon dioxide and methane ("Atmospheric Science" 2001). Gardner and colleagues (2001) report measurements of the temperature, iron density, and altitude of polar mesospheric clouds (PMCs) from an airplane. Results are reported in *Geophysical Research Letters* (Gardner et al. 2001).

PSCs have been described as "Nacreous clouds resembling giant abalone shells floating in the sky" (Tolbert and Toon, 2001, 61). Occasional stratospheric clouds have been reported in Scandinavia for a century, and Edward Wilson noted them during Robert Falcon Scott's 1901 Antarctic expedition. Sometimes the clouds shine with green and orange shades at sunrise and sunset (Tolbert and Toon, 2001, 61).

PSCs remained largely an atmospheric curiosity until the discovery of widespread ozone depletion over the Antarctic during the middle 1980s. Scientists surmised that the ozone loss was occurring in the only place where the stratosphere is cold enough to produce clouds. The clouds form during the springtime, when sunshine is available, and their coverage seems related to rising levels of chlorinated molecules, including CFCs and other synthetics. Observers have noted "unprecedented concentrations of reactive chlorine in conjunction with severe ozone loss" (Tolbert and Toon 2001, 61). Ozone loss during the winter of 1999–2000 was exceptionally severe, as record cold stratospheric temperatures produced abundant polar stratospheric cloudiness.


## THE EVOLUTIONARY NATURE OF THE SCIENCE

The relationship between CFCs, ozone depletion, and global warming illustrates the evolutionary nature of the science of atmospheric chemistry. Industry invents the chemicals and puts them to work, in the meantime changing nature's chemistry in ways that science detects and describes only decades later, often after substantial ecological damage has been done.

A theoretical possibility that global warming could hasten ozone depletion was suggested by mathematical model results reported in the November 19, 1992, issue of *Nature*. J. Austin, N. Butchart, and K. P. Shine (1992), climate modelers from the British Meteorological Office and Reading University, ran a model simulating conditions in the stratosphere in a world where atmospheric carbon dioxide concentrations had been allowed to double, compared to preindustrial levels, which is the world in which we will be living in within fifty years at present growth rates.

## OZONE LOSS CONFIRMED IN THE ARCTIC

Scientists have been looking for reasons that the ozone shield has failed to replenish itself after the international ban on production of CFCs. During the middle 1990s, scientists began to model a relationship between global warming and ozone depletion.

A team led by Drew Shindell, an atmospheric chemist at the NASA Goddard Institute for Space Studies in New York City, created the first atmospheric simulation to include ozone chemistry. The team found that the greenhouse effect was responsible not only for heating the lower atmosphere, but also for cooling the upper atmosphere. The cooling poses problems for ozone molecules, which are most unstable at low temperatures. Based on the team's model, the buildup of greenhouse gases could chill the high atmosphere near the poles by as much as 8°C to 10°C. The model predicted that maximum ozone loss would occur between the years 2010 and 2019 (Shindell, Rind, and Lonergan 1998, 589).

Stratospheric ozone loss in the Arctic was confirmed during the winter of 1999–2000. An international group of researchers found cumulative ozone losses of more than 60 percent at around 18 kilometers (11 miles) above the Arctic between January and March 2000. "These are among the largest chemical losses at this altitude observed during the last 10 years," according to a statement of the European Union, a main sponsor of the research (British Broadcasting 2000). European Union spokeswoman Piia Huusela said the report did not point to a hole in the ozone layer like the one that has opened over the Antarctic, but a weakening of ozone content in the stratosphere. Ozone amounts over the Arctic today are now said to be 15 percent below the pre-1976 average. "This is not a hole in the ozone layer," said Huusela. "We are not even close to a hole, but it is nevertheless alarming" (British Broadcasting 2000).

Jonathan Shanklin, of the British Antarctic Survey and one of the three scientists credited with discovering the ozone hole over Antarctica, warned in late October 2000, that global warming eventually threatens to create an ozone "hole" over the Arctic that eventually may rival the Antarctic "hole" in size and severity. Shanklin told BBC Radio 4's "Costing the Earth" program that Arctic ozone depletion could affect the United Kingdom, bathing it in higher levels of cancer-causing ultraviolet radiation. Shanklin said that the buildup of greenhouse gases is trapping the sun's heat, making the Earth warmer, which has the effect of making the higher ozone layer colder and increasing the catalytic action of ozone-killing chemicals that concentrate over the poles (Nuttall 2000).

Shanklin, who, during 1985 first detected substantial stratospheric ozone depletion over the Antarctic with Joseph Farman and Brian Gardiner of the British Antarctic Survey, said:

The atmosphere is changing, and one of the key changes is that the ozone layer is getting colder. It's getting colder because of the greenhouse gases that are being liberated by all the emissions we have at the surface. And when it gets colder, particularly during the winter, we can get clouds actually forming in the ozone layer, and these clouds are the key factor. Chemistry can take place on them that activates the chlorine and makes it very much easier for it to destroy the ozone. We think that within the next 20 years we're likely to see an ozone hole perhaps as big as the present one over Antarctica, but over the North Pole. (Kirby 2000)

While most of the area covered by the Antarctic ozone hole is uninhabited by human beings, a similar Arctic hole would affect parts of densely populated Europe, Asia, and North America. Following the winter of 1999–2000's record ozone loss in the Arctic, a European Space Agency satellite detected evidence that ozone levels above Great Britain, Belgium, the Netherlands, and Scandinavia for short periods "were nearly as low as those normally found in the Antarctic" (Baker 2000, 38).

The World Meteorological Organization supported Shanklin's assessment:

Chemicals that result in ozone destruction are no longer increasing in the stratosphere, as the international controls on ozone-depleting chemicals continue to work. However, the continued general decrease of ozone in the lower stratosphere and the global increase in greenhouse gases are now believed to result in lower temperatures in the lower stratosphere. These decreases in temperature could expand the period of intense ozone loss. (Kirby 2000)

## THE IMPORTANCE OF TEMPERATURE SENSITIVITY

The extreme temperature sensitivity of these reactions has tremendous consequences for Arctic ozone. A small cooling in the Arctic stratosphere may provoke a much greater ozone loss. In some cases, a 5°C decline in temperatures can multiply ozone loss ten times.

Ozone depletion itself intensifies stratospheric cooling by reducing the upper atmosphere's ability to absorb incoming ultraviolet solar radiation. The amount of supercooled stratospheric territory increased dramatically during the 1990s.

In 1998 the region of severely depleted ozone ("the ozone hole") over the Antarctic reached a record size roughly the size of the continental United States (it was three times that size two years later). Some researchers have come to the conclusion that, as Richard A. Kerr described in *Science:* "Unprecedented stratospheric cold is driving the extreme ozone destruction. . . . Some of the high-altitude chill . . . may be a counterintuitive effect of the accumulating greenhouse gases that seem to be warming the lower atmosphere. The colder the stratosphere, the greater the destruction of ozone by CFCs" (Kerr 1998, 391).

"The chemical reactions responsible for stratospheric ozone depletion are extremely sensitive to temperature," Shindell said. He continued: "Greenhouse gases warm the Earth's surface but cool the stratosphere radiatively, and therefore affect ozone depletion." By the decade 2010 to 2019, Shindell's team expects ozone losses in the Arctic to peak at two-thirds of the ozone column, or roughly the same ozone loss observed in Antarctica by the early 1990s. "The severity and duration of the Antarctic ozone hole are also expected to increase because of greenhouse-gas-induced stratospheric cooling over the coming decades," Shindell has asserted (Shindell, Rind, and Lonergan 1998, 589).

During the middle 1990s, scientists began to detect ozone depletion in the Arctic after a decade of measuring accelerating ozone loss over the Antarctic. By the year 2000, the ozone shield over the Arctic had thinned to about half its previous density during March and April. Ozone depletion over the Arctic reaches its height in late winter and early spring, as the sun rises after the midwinter night. Solar radiation triggers reactions between ozone in the stratosphere and chemicals containing chlorine or bromine. These chemical reactions occur most quickly on the surfaces of ice particles that form stratospheric clouds at temperatures under minus 80°C (minus 107°F).

## NEAR-SURFACE WARMING ABETS
## STRATOSPHERIC COOLING

Space-based temperature measurements of the Earth's lower stratosphere, a layer of the atmosphere from about seventeen kilometers to twenty-two kilometers (roughly ten to fourteen miles) above the surface, indicate record cold at that level as record surface warmth was reported during the 1990s. Roy Spencer of NASA and John Christy of the University of Alabama at Huntsville and the Global Hydrology and Climate Center obtained temperature measurements of layers within the entire atmosphere of the Earth from space, using microwave sensors aboard several polar-orbiting weather satellites. They found that, despite significant short-lived warming after the eruptions of El Chichón in Mexico in 1982 and Mt. Pinatubo in the Philippines in 1991, the stratosphere has been cooling steadily during the past fifteen years. During the winter of 1999–2000, temperatures in the stratosphere over the Arctic were recorded at minus 118°F, the lowest on record.

As Dennis L. Hartmann and colleagues (2000) explained,

The pattern of climate trends during the past few decades is marked by rapid cooling and ozone depletion in the polar lower stratosphere of both hemispheres, coupled with an increasing strength of the wintertime westerly polar vortex and a poleward shift of the westerly wind belt at the Earth's surface. . . . [I]nternal dynamical feedbacks within the climate system . . . can show a large response to rather modest external forcing. . . . Strong synergistic interactions between stratospheric ozone depletion and greenhouse warming are possible. These interactions may be responsible for the pronounced changes in tropospheric and stratospheric climate observed during the past few decades. If these trends continue, they could have important implications for the climate of the twenty-first century. (1412)

Ozone depletion has been measured only for a few decades, and so researchers caution that they are not entirely certain that rapid warming at the surface is not being caused by natural variations in climate that are powerfully influenced by the interactions of oceans and atmosphere. "However," Hartmann and colleagues conclude, "It seems quite likely that they are at least in part human-induced." Hartmann and associates also have raised the possibility that the poleward shift in westerly winds may be accelerating melting of the Arctic ice cap, part of what they contend may be a "transition of the Arctic Ocean to an ice-free state during the twenty-first century" (Hartmann et al., 2000, 1412). A continued northward shift in polar westerly winds in winter also could portend additional warming over the land masses of North America and Eurasia.

The connection between global warming, a cooling stratosphere, and depletion of stratospheric ozone was confirmed in April 2000, with release of a lengthy report by more than three hundred NASA researchers as well as several European, Japanese, and Canadian scientists. The report found that while ozone depletion may have stabilized over the Antarctic, ozone levels north of the Arctic circle were still falling, in large part because the stratosphere has cooled as the troposphere has warmed. The ozone level over some parts of the Arctic was 60 percent lower during the winter of 2000 than during the winter of 1999.

In addition, scientists learned that, in coming years as winter ends, the ozone-depleted atmosphere may tend to spread southward over heavily populated areas of North America and Eurasia. "The largest ultraviolet increases from all of this are predicted to be in the mid-latitudes of the United States," said University of Colorado atmospheric scientist Brian Toon. "It affects us much more than the Antarctic," said Seth Borenstein (2000).


## PROJECT SOLVE

NASA's contribution to the Arctic ozone survey has been dubbed SOLVE [SAGE (Stratospheric Aerosol and Gas Experiment) III Ozone Loss and Validation Experiment]. SOLVE used satellites, aircraft, balloons, and ground-based instruments between November 1999 and March 2000 to document changes in the Arctic ozone shield. Scientists also gathered ozone-related data using the Russian Meteor-3 satellite, which was used to measure the vertical structure of aerosols, ozone, water vapor, and other trace gases in the Arctic upper troposphere and stratosphere. This information is being used by more than two hundred scientists and support staff from the United States, Canada, Europe, Russia, and Japan.

SOLVE and its Canadian, European, Russian, and Japanese counterparts is the largest field measurement campaign devoted to measure ozone amounts and changes in the Arctic upper atmosphere. Researchers examined the processes that control ozone levels at mid- to high latitudes during the Arctic winter between

November 1999 and March 2000. All this effort is being directed toward understanding why the ozone column over the Arctic continues to deteriorate despite the banning of the main culprit in stratospheric ozone loss, CFCs. Accelerating ozone loss has been measured since the early 1990s in the Arctic.

The project's main field office was established above the Arctic Circle at the airport in Kiruna, Sweden. Arena Arctica, a large hangar especially built for research, housed the aircraft and many of the project's scientific instruments. Balloons were launched from Esrange, a launch facility near Kiruna, where wintertime conditions can be very severe, with temperatures falling below minus 50°F.

Kiruna is an optimal site for sampling Polar Stratospheric Clouds because the stratosphere above Kiruna is usually quite cold, and local meteorological conditions include mountain-forced waves, which create even colder conditions, leading to PSC formation. Kiruna also is near the climatological average coldest region in the Arctic. On average, the coldest point in the Arctic lower stratosphere is located over Spitzbergen Island, a short flight from Kiruna.

Ross Salawitch, a research scientist at NASA's Jet Propulsion Laboratory in Pasadena, California, said that if the pattern of extended cold temperatures in the Arctic stratosphere continues, ozone loss over the region could become "pretty disastrous" (Scientists Report 2000, 3-A). Salawitch said that the new data has "really solidified our view" that the ozone layer is sensitive not only to ozone-destroying chemicals but also to temperature (Stevens 2000, A19).

"The temperature of the stratosphere is controlled by the weather that will come up from the lower atmosphere," said Paul Newman, another scientist who is taking part in the Arctic ozone project. "If we have a very active stratosphere we tend to have warm years; when stratosphere weather is quiescent we have cold years" (Connor 2000, 5). New research indicates that global warming will continue to cool the stratosphere, making ozone destruction more prevalent even as the volume of CFCs in the stratosphere is slowly reduced. "One year does not prove a case," said Newman, who works at NASA's Goddard Space Flight Center in Greenbelt, Maryland. "But we have seen quite a few years lately in which the stratosphere has been colder than normal" (Connor 2000, 5).

"We do know that if the temperatures in the stratosphere are lower, more clouds will form and persist, and these conditions will lead to more ozone loss," said Michelle Santee, an atmospheric scientist at NASA's Jet Propulsion Laboratory in Pasadena and coauthor of a study on the subject in the May 26, 2000, issue of *Science* (McFarling 2000, A20). The anticipated increase in cloudiness over the Arctic could itself become a factor in ozone depletion. The clouds, formed from condensed nitric acid and water, tend to increase snowfall, which accelerates depletion of stratospheric nitrogen. The nitrogen (which would have acted to stem some of the ozone loss had it remained in the stratosphere) is carried to the surface as snow.

## PROJECTIONS OF STRATOSPHERIC OZONE RECOVERY

Like all weather forecasts, the future of stratospheric ozone recovery is open to debate. One of the more optimistic forecasts states that severe depletion in the Southern Hemisphere's ozone layer will start shrinking within a decade and should close completely within the next fifty years. "Data from the Cape Grimm monitoring station in Tasmania show that levels of CFCs in the lower atmosphere are starting to decline for the first time since scientists from the British Antarctic Survey discovered the ozone hole in 1985. A new mathematical model, the most accurate yet devised, suggests that there will be a similar decline in the stratosphere over the next decade, leading to a recovery in levels of ozone" (Henderson 2000).

The report contains several caveats. "The dramatic recovery could, however, be slowed by as much as 30 years by global warming or by severe volcanic eruptions, according to the meeting in Argentina of the Stratospheric Processes and Their Role in Climate panel, which is a project of the World Climate Program. It will also depend on continued efforts of the global community to keep ozone emissions low. The hole could also grow slightly over the next five years before recovery begins" (Henderson 2000).

Alan O'Neill, the director of the Centre for Global Atmospheric Modelling, University of Reading, and chairman of the panel, said that the news was a "triumph" for global cooperation (Henderson 2000). The success could be attributed to the 1987 Montreal Protocol, in which most governments pledged to reduce their use of CFCs, he said.

Worldwide production of CFCs has declined sharply. The United States has cut its annual output of ozone-depleting chemicals from

306,000 ozone-depletion potential tons (ODP tons) to 2,500 since the Montreal Protocol was negotiated. The twelve nations that were then members of the European Union have reduced their use from 301,000 to 4,300 ODP tons, while Japan has cut its output from 118,000 ODP tons to zero (Henderson 2000).

Sherwood Rowland of the University of California at Irvine, who shared the 1995 Nobel Prize for Chemistry for his part in describing the chemistry of stratospheric ozone depletion, believes that the effect of global warming on ozone depletion should be short-lived. "The [ozone depletion] story is approaching closure, and that's very satisfying," Rowland said (Schrope 2000, 627).

Drew Shindell agreed with Rowland, saying that even though scientists are beginning to understand how global warming could delay ozone-shield recovery, "The agreement to limit production [of CFCs] has been an unqualified success. The science was listened to, the policy-makers did something, and it actually worked" (Schrope 2000, 627).

## "ROCKS" IN THE STRATOSPHERE

During the year 2001, new scientific knowledge describing the relationship of global warming and stratospheric ozone depletion began to be published in leading journals. In *Science,* a team led by A. Tabazadeh described a relationship between stratospheric cooling, denitrification, and global warming. Tabazadeh and colleagues (2001) wrote:

Homogeneous freezing of nitric acid hydrate particles can produce a polar freezing belt in either hemisphere that can cause denitrification. . . . A 4 kelvin decrease in the temperature of the Arctic stratosphere due to anthropogenic and/or natural effects can trigger the occurrence of widespread severe denitrification. Ozone loss is amplified in a denitrified stratosphere, so the effects of falling temperatures in promoting denitrification must be considered in assessment studies of ozone-recovery trends. (2591)

At about the same time, a team of atmospheric scientists led by atmospheric chemist David W. Fahey (of the National Oceanic and Atmospheric Administration's office in Boulder, Colorado) discovered large particles inside stratospheric clouds over the Arctic that could delay the healing of the Earth's protective ozone layer. The team found large nitric-acid-containing particles that could delay the recovery and make the ozone layers over both poles more vulnerable to climate change.

Fahey led a team of twenty-seven researchers that included scientists from the National Oceanic and Atmospheric Administration in Boulder, the University of Colorado, the National Center for Atmospheric Research in Boulder, and the University of Denver. The scientists described their findings in the February 9, 2001, edition of the journal *Science* (Fahey et al. 2001). "It's a major puzzle piece in the process by which ozone comes to be destroyed," Fahey said of the discovery, which occurred during a January 2000 flight over the Arctic in the ER-2, NASA's version of the U-2 spy plane (Erickson 2001, 37-A). According to Richard Kerr, a machine on the aircraft that was measuring nitrogen-containing gases "coughed out what looked like disastrous noise" (Kerr 2001, 962). The "noise" turned out to be very large particles (compared to other masses in Arctic clouds) containing a form of nitric acid ($HNO_3$), previously unknown to science. The particles averaged 3,000 times the size of other atmospheric particles in the stratosphere.

Each winter in the stratosphere over the poles, water and nitric acid condense to form clouds that unleash chlorine and bromine, which degrade ozone. Later in the winter, nitrogen compounds help shut down the destruction. Fahey's team found previously unknown nitric acid particles that remove nitrogen, allowing the destruction to continue. They nicknamed them "rocks" because they are so much larger than any other type of particle in PSCs.

The "rocks" form during the polar winter, when stratospheric temperatures are at their lowest. Cooling of the stratosphere compelled by the retention of heat near the surface of the Earth may cause more "rocks" to form, accelerating ozone depletion. "If it gets colder and you get more 'rocks,' the depletion period is going to last longer. The chlorine can continue to eat ozone," said Paul Newman, an atmospheric physicist at NASA's Goddard Space Flight Center in Maryland (Erickson 2001, 37-A). "What [they] got is really outstanding," Newman said of the findings by Fahey's team. "This mechanism that we now understand really will help us be able to more precisely predict what's going to happen in the future."

No one yet has even hazarded so much as a guess about how the "rocks" form. These PSC particles remove nitrogen from the atmosphere that would otherwise "tie up chlorine and bromine in inactive, harmless forms" through denitrification (Kerr 2001, 963).

The "rocks" also "provide surfaces where chlorine and bromine can be liberated from their inactive forms to enter their ozone-destroying forms" (Kerr 2001, 963). In addition, the large size of the PSC "rocks" causes them to fall more quickly than other particles, removing even more nitrogen from the stratosphere. Through all these mechanisms, according to Fahey and his colleagues, (2001, 1026) the PSC "rocks" "have significant potential to denitrify the lower stratosphere."

Fahey and his colleagues (2001) concluded:

Arctic ozone abundances will remain vulnerable to increased winter/spring loss in the coming decades as anthropogenic chlorine compounds are gradually removed from the atmosphere, particularly if rising concentrations of greenhouse gases induce cooling in the polar vortex and trends of increasing water vapor continue in the lower stratosphere. Both effects increase the extent of PSC formation and, thereby, denitrification and the lifetime of active chlorine. The role of denitrification in these future scenarios is likely quite important. (1030)

Fahey and his colleagues estimate that ozone depletion in the Arctic stratosphere may not reach its peak until the year 2070, even with a steady decline in CFC levels.

## THE "FRIO BANDITOS"

Ozone-consuming CFCs continue to be used in older air conditioners in the United States. Because manufacture of CFCs is illegal in the United States, a flourishing smuggling trade has developed from Mexico. Bert Ammons of Stuart, Florida, for example, pled guilty to violating the Clean Air Act when he attempted to smuggle ninety thirty-pound cylinders of CFC-12, also known by its trade name, Freon, in false compartments on his forty-one-foot boat, *Sierra*. According to Environmental Protection Agency officials, Ammons had planned to distribute $68,000 worth of smuggled CFCs to auto repair shops around Fort Lauderdale. The crimes of which Ammons was accused carry a maximum of five years in prison and $250,000 in fines (Baker 2000, 34).

In Mexico and China (among other developing nations), CFC-12 can be bought for $1 or $2 a pound and resold in the United States for $20 to $25 a pound. Why the huge domestic markup? It's a simple matter of supply and demand (Baker 2000, 34). Millions of pieces of equipment that use CFCs are still in service, including automobiles built before 1994, air conditioners, and other refrigeration equipment.

Between 1994 and 1997, at least 6,367 tons of CFC-12 and 24 tons of CFC-113 (used as a fire suppressant) were smuggled across the U.S. border. According to an unnamed official in the EPA's Criminal Enforcement Division, "Illegal CFCs rank close to cocaine as some of the most profitable contraband coming across the U.S. border" (Baker 2000, 34). Smugglers of Freon over the Mexican border have come to be called "frio banditos" (Baker 2000, 36).

By the late 1990s, CFC smuggling had become a multimillion dollar business, big enough for the Justice Department and the Environmental Protection Agency to make a priority of prosecuting it. During January 2003, for example, a ring of CFC smugglers that had collected more than $6 million in profits between 1996 and 1998 were sentenced. Barry Himes, the leader of the ring, was ordered to spend up to six and a half years in prison. Himes, who also forfeited a $3 million mansion, a BMW sedan, and a three-carat diamond ring, also was ordered to pay $1.8 million in restitution. Coconspirator John Mucha was sentenced to up to four years in prison and $1.2 million in restitution. Ten defendants were charged as part of this particular smuggling operation. In all, by early 2003, 114 people had been charged with smuggling CFCs into the United States.

## SHOWDOWN OVER METHYL BROMIDE

Another threat to stratospheric ozone is methyl bromide, which is used in the United States primarily by California strawberry and Florida tomato growers. A showdown over use of this chemical has been ongoing in Congress, where legislation introduced by California congressman Richard Pombo (the Methyl Bromide Fairness Act) would push back the U.S. phase-out date of the chemical to 2015—the year developing countries are required to stop production and consumption of the chemical (Baker 2000, 34).

"Due to its acute toxicity," one observer has written, "methyl bromide is already banned in several countries, including the Netherlands and Canada. For years, environmentalists and health officials in the U.S. (which uses 40 percent of the world's methyl bromide) have called for stricter regulation of this pesticide, especially in agricultural areas such as California's Ventura County, where children and farm workers are at risk. Since 1982, nearly 500 poisonings linked to methyl bromide have occurred in California, 19 of them fatal" (Baker 2000, 34).

Azadeh Tabazadeh, an atmospheric chemist at the NASA Ames Research Center, Mountain View, California, said, "the bromide in methyl bromine is a much better catalyst for ozone destruction than chlorine." She added: "And just because we've reduced the amount of chlorine in the atmosphere doesn't mean that the level of bromine is also going down. That's why compounds like methyl bromide need to be regulated" (Baker 2000, 37). The U.S. EPA classifies methyl bromide as a Class 1 ozone-depleting substance to be phased out under the provisions of the Clear Air Act.

Unregulated emissions of methyl bromide have increased, while manufacture of CFCs has all but ended in the United States. A report of the United Nations Environmental Program suggests that methyl bromide production and use may be the single most important variable in ozone depletion during the next several decades.

## EROSION OF THE ATMOSPHERE'S NATURAL CLEANSERS

A sharp drop in atmospheric concentrations of hydroxyl radicals, chemicals that naturally purge the air of contaminants, has been detected by scientists. The biggest drop has been noted in the Northern Hemisphere, site of most of humanity's industrial infrastructure, and now home to an aerosol mist that may be accelerating the chemicals' demise. Hydroxyls oxidize pollutants, making them soluble in water so that they wash away as part of liquid precipitation.

These molecules' continued decline could spread smog and accelerate atmospheric accumulation of greenhouse gases, as the Earth loses one of its principal defenses against several polluting compounds, including carbon monoxide, methane, and sulfur dioxide. During the last twenty-two years concentrations of the molecule have decreased an average of 10 percent worldwide, the study found (Wang and Prinn 1998). One of the study's authors is Ronald G. Prinn, chair of the Department of Earth, Atmospheric, and Planetary Sciences at the Massachusetts Institute of Technology.

Hydroxyl radicals "are created as ultraviolet light knocks hydrogen atoms from water molecules in air in the presence of ozone, a highly reactive form of oxygen" (Revkin 2001). The radicals vanish almost as quickly as they are created, usually in less than a second, as they react chemically with several air pollutants, including carbon monoxide, methane, and sulfur dioxide. One of many possible influences, atmospheric scientists say, is an increase in aerosol haze, a common by-product of fossil-fuel combustion, that could block ultraviolet light and impede the reaction that creates the molecules (Revkin 2001). Hydroxyl radicals may purge more than half the sulfur dioxide added to the air by industry, volcanoes, and other sources.

Ralph J. Cicerone, an atmospheric chemist at the University of California at Irvine, said about these new findings that "This is a terrifically important question because hydroxyl radicals are the central chemical in the lower atmosphere for processing everything," he said. "For 25 years, people have been struggling to measure [them]" (Revkin 2001).

Because hydroxyl radicals' lives are so short, scientists often must study them indirectly, usually by examining changes in concentrations of methyl chloroform, a long-lived solvent in the air that can be neutralized only by the hydroxyl radicals. Methyl chloroform's manufacture was banned during the 1990s because it also contributes to the destruction of stratospheric ozone.

Stephen A. Montzka, a research chemist for the National Oceanic and Atmospheric Administration, reflected on the fact that the scientists are using proxy measurements of the radicals that may not be reliable. However, he said that Prinn's work has produced "the best barometer of hydroxyl radicals that we have, but there are still big potential sources of error" (Revkin 2001).

"This one molecule is very, very important. It is the critical cleaning chemical for the atmosphere," said Prinn, who led a thirteen-member research team for the study. "If this free-radical [molecule] is decreasing, it could add to global warming" (Polakovic 2001, A1). Losses of the chemical, a hydroxyl radical, "are slight so far and are not currently cause for alarm" (Polakovic 2001, A1). "If this change has happened, it is a slight change. It's significant if it's the beginning of a trend. That would be a warning," said Sherwood Rowland, a chemist at the University of California at Irvine (Polakovic 2001, A1).

"We are seeing for the first time that the oxidizing capacity of the atmosphere may have been reduced in recent years," said Ray F. Weiss, professor of geochemistry at the Scripps Institution of

Oceanography (Polakovic 2001, A1). These molecules also are especially efficient neutralizers of hydrocarbons from paints, solvents, and petroleum that cause smog and cancer. About 18 million tons of hydrocarbons are released into the atmosphere annually in the United States (Polakovic 2001). "Eventually, the atmosphere will get to the point where it will be taxed beyond its ability to clean itself," said Chris Cantrell, atmospheric chemist at the National Center for Atmospheric Research in Colorado (Polakovic 2001, A1). Human emissions of various pollutants may overwhelm the atmosphere's remaining defenses within fifty years, according to California Institute of Technology scientist John H. Seinfeld (Polakovic 2001, A-1).

Prinn and colleagues acknowledge a wide margin of error in their data. They assert, more specifically, that "the actual loss of cleansing molecules could be 24 percent greater than they have predicted, or much less" (Polakovic 2001, A1). "Unless we cut the amount of pollution we produce, we could end up facing a horrendous dilemma: Mend the ozone layer and suffocate in smog, or leave the ozone layer damaged and let hydroxyl [radicals] keep cleaning up our atmosphere. Either option is disastrous," wrote Jeremy Webb, editor of the *New Scientist* (Freeman 2001, 3).

The *New Scientist* painted a doomsday picture of the year 2050— a world in which asthma has become the number-one killer of people thirty years of age, "where Japan chokes in the sulfur dioxide fumes from China, and where famine takes hold across Russia due to crop failure" (Freeman 2001, 3). By 2070, this *New Scientist* report "imagines that the fortunate few will be taking refuge in great city domes, abandoning billions more to live outside amid the contamination where life expectancy is reduced to 30 years" (Freeman 2001, 3).

## GLOBAL WARMING, OZONE DEPLETION, AND THE DEVASTATION OF THE BLUE WHALE

Global warming and ozone depletion interface in many ways. Consider the impending demise of the blue whale, the largest animal ever to live on Earth. The blue whale is endangered because of global warming as well as ozone depletion, according to the World Wildlife Fund. Populations have fallen from an estimated 300,000 whales in 1900 (250,000 of them in waters around Antarctica) to fewer than 5,000 by the year 2000. The blue whale's main food source, shrimp-like krill, are dying because of climate change. The whales are going hungry after their populations had been reduced by decades of aggressive whaling.

Krill feed on microscopic marine algae frozen in polar sea ice, which are released each summer when the ice melts. The World Wildlife Fund asserts that as the Earth's temperatures have increased in recent decades, sea ice has diminished enough to materially reduce krill populations. The WWF report also contends that sharp reductions in stratospheric ozone density are allowing ultraviolet radiation to cause DNA damage and excess mortality in krill. Stuart Chapman, a whale specialist for WWF, said, "If this decline continues . . . it could lead to the extinction of the Antarctic blue whale" (Urquhart 2001, 7).

## THE EVOLVING COMPLEXITY OF STRATOSPHERIC OZONE CHEMISTRY

By 2002 most atmospheric chemists were recognizing that healing stratospheric ozone was not as simple as outlawing manufacture of chemicals, such as CFCs, that increase levels of chlorine in the upper atmosphere. A Canadian study released in early 2002 cast doubt on earlier optimistic prospects that stratospheric ozone levels would steadily increase throughout the twenty-first century as levels of chlorinated compounds declined. "The more we know, the more we realize we don't know," said Jack McConnell, an atmospheric science professor at York University in Ontario, Canada (Calamai 2002, A8). The Canadian study projected a decline in stratospheric ozone levels over Canada throughout the century, leading to a rise in some skin cancers of about 10 percent (Calamai 2002, A8).

The Canadian study cited new research by Australia's national scientific agency CSIRO that examined the relationship between stratospheric ozone levels and two key greenhouse gases, nitrous oxide and methane. Nitrous oxide, which remains in the atmosphere about 120 years, comes from sources that are difficult to control, such as agricultural soils and vehicle emissions, so a continued buildup in the atmosphere is almost certain. As it slowly breaks down, nitrous oxide destroys ozone (Calamai 2002, A8).

According to an account in the *Toronto Star* by Peter Calamai,

A CSIRO computer model described in the journal *Geophysical Research Letters* forecasts that this combination will keep levels of stratospheric ozone depressed throughout this century, despite the global ban on the chlorine chemicals responsible for the ozone holes over the North and South poles. The most optimistic projection is that ozone levels at the end of the century would be roughly where they are now—5 percent below those in 1980 when the problem began. The model also forecasts that reducing methane emissions, as Toronto has done at landfill sites—actually makes matters worse for protective ozone and could drive the levels in 2100 down to 9 percent below 1980 levels. (Calamai 2002, A8)

Methane, which accelerates global warming, helps to protect against ultraviolet radiation because it produces ozone as it breaks down chemically. Current global climate change strategy focuses on pushing down methane levels while letting nitrous oxide levels rise, which probably will cause ozone levels to fall.

The complexity of ozone-depletion chemistry indicates the risks inherent in any attempt to manipulate the atmosphere for one outcome without regard for others. "You have to look at all these chemicals and see how they interact and evolve over time," said Tom McElroy, an ozone specialist with the Meteorological Service of Canada (Calamai 2002, A8).

Yet another problem for stratospheric ozone may be wildfires and slash-and-burn agriculture in the tropics, major factors in a doubling of moisture content in the stratosphere during the last half of the twentieth century. "In the stratosphere, there has been a cooling trend that is now believed to be contributing to milder winters in parts of the northern hemisphere," said Steven Sherwood, assistant professor of geology and geophysics at Yale University. "The cooling is caused as much by the increased humidity as by carbon dioxide" ("Biomass Burning" 2002). "Higher humidity also helps catalyze the destruction of the ozone layer," Sherwood wrote in a *Science* journal article. "More aerosols lead to smaller ice crystals and more water vapor entering the stratosphere" (Sherwood 2002, 1272).

Cooling in the stratosphere causes changes to the jet stream that produce milder winters in North America and Europe. By contrast, harsher winters result in the Arctic ("Biomass Burning" 2002). Sherwood said that about half the increased humidity in the stratosphere has been attributed to methane oxidation. It was not known, however, what has caused the remaining additional moisture. Sherwood said that "Aerosols are smoke from burning. They fluctuate seasonally and geographically. Over decades there have been increases linked to population growth" ("Biomass Burning" 2002).

"Volcanic aerosols have been linked by A. Tabazadeh and her colleagues to loss of ozone over the Arctic, especially in the springtime, due to denitrification and chlorine activation. This is related to the fact that volcanic eruptions produce "clouds that have greater surface area than typical arctic polar stratospheric clouds" (Tabazadeh et al., 2002, 2609). The authors state that "Chemical processing on volcanic aerosols over a 10-kilometer altitude range could increase the current levels of springtime column ozone loss up to 70 percent independent of denitrification" (Tabazadeh et al., 2002, 2609). A cooling of the lower stratosphere due to global warming near the surface may aggravate this loss.

Unlike the Antarctic, where the stratosphere is consistently cold, stratospheric "weather" in the Arctic varies greatly. The general trend has been toward colder stratospheric temperatures in part because of global warming near the surface. According to Tabazadeh and her colleagues, "Significant ozone loss in the Arctic occurs only in cold winters, and volcanoes can substantially increase this loss by enhancing the spatial scales over which ozone molecules can get destroyed in the stratosphere" (Tabazadeh et al., 2002, 2609). Because they increase the risk of ozone loss in the Arctic (in conjunction with colder than usual winters) volcanic activity will need to be monitored especially closely during the next thirty years, "while anthropogenic chlorine levels are still sufficiently high (about 3 p.p.b.) to cause severe ozone depletion" (Tabazadeh et al., 2002, 2609). Models indicate that early rapid growth in the Antarctic ozone "hole" in the early 1980s may have been influenced by a higher than usual level of volcanic activity, according to Tabazadeh and her colleagues.

According to an account by the Environment News Service, "Large volcanic eruptions pump sulfur compounds into the Earth's atmosphere. These compounds form sulfuric acid clouds similar to polar stratospheric clouds made of nitric acid and water. The clouds of nitric acid and water form in the upper atmosphere during very cold conditions and play a major part in the destruction of ozone over Earth's poles" ("Volcanic Eruptions" 2002).

After eruptions, volcanic sulfuric acid clouds would boost the ozone-destroying power of polar stratospheric clouds, the researchers said. "Volcanic aerosols also can cause ozone destruction at warmer temperatures than polar stratospheric clouds, and this would expand the area of ozone destruction over more populated

areas," Tabazadeh said. "Nearly one-third of the total ozone depletion could be a result of volcanic aerosol effects at altitudes below
about 17 kilometers (11.5 miles)," the researchers wrote. "Climate
change combined with after-effects of large volcanic eruptions will
contribute to more ozone loss over both poles," Tabazadeh concluded. "This research proves that ozone recovery is more complex
than originally thought" ("Volcanic Eruptions" 2002).

The authors conclude that a period of higher than usual volcanic
activity coupled with colder than usual temperatures in the stratosphere could cause ozone levels there to fall to "values measured
inside the Antarctic 'ozone hole.'" The researchers place the probability of such a combination of circumstances at about 15 percent
during the next thirty years—50 percent chance of a "cold" year
times 0.31 percent chance of major volcanic activity, or what they
call a "volcanic cloudy year" (Tabazadeh et al., 2002, 2612). "Therefore," they write, "It is possible for about five Arctic winters in the
next three decades to be cold and volcanic."

"A 'volcanic ozone hole' is likely to occur over the Arctic within the
next 30 years," said Azadeh Tabazadeh, lead author of the paper
and a scientist at NASA's Ames Research Center. "If a period of high
volcanic activity coincides with a series of cold Arctic winters, then
a springtime Arctic ozone hole may reappear for a number of con-
secutive years, resembling the pattern seen in the Antarctic every
spring since the 1980s," Tabazadeh added (Volcanic, 2002).

## ENDURING CONCERNS

The synergies of climate change—such as the interstice of stratospheric ozone depletion and global warming—may compound the
effects of any single phenomenon. As the impending extinction of
the Blue Whale illustrates, the ecological toll of one anthropogenic
process may be compounded by the impact of another. For example,
rising levels of methane in the stratosphere could increase levels of
water vapor there, allowing, when temperatures are cold enough,
larger numbers of PSCs, and thus more ozone depletion.

All the while, as science has discovered new ways to describe
these synergies, climatologists have been sharing the disquieting
notion that small shifts in global temperature could lead to sudden
and abrupt climate changes. The history of atmospheric chemistry
during the last few decades has been one of surprises. The major

provocation of the ozone hole was chlorofluorocarbons. According to Paul Crutzen, an atmospheric chemist working in Germany who shared a Nobel Prize for detecting ozone depletion over Antarctica, "Had chemists earlier in the century decided to use bromine instead of chlorine to produce coolants, a mere quirk of chemistry—the ozone hole would have been far larger, occurred all year and severely affected life" (McFarling 2001, A1). "Avoiding that was just luck," he said, noting that no scientist had predicted the scope of ozone depletion. "We missed something very important. There may be more of these things around the corner" (McFarling 2001, A1).

During the fall of 2002, nature threw scientists studying stratospheric ozone loss a curve. The area of severely depleted ozone over Antarctica shrunk and split into two pieces during September 2002. In 2002, the depleted area was smaller than at any time since 1988, a shrinkage that scientists attributed to warmer-than-usual temperatures in the stratosphere caused by atmospheric agitation.

"The Southern Hemisphere's stratosphere was unusually disturbed this year," said Craig Long, meteorologist at NOAA's Climate Prediction Center. The size of the ozone-depleted area shrunk to roughly 15 million square kilometers from its record size of 24 million during September 2001. In addition, "This breakdown [of the ozone-depleted area] is occurring exceptionally early in the year, about two months earlier than normal," said Henk Eskes, a senior scientist at the Royal Netherlands Meteorological Institute. "The depth of the ozone hole this year also is unusually small, about half that recorded in 2001" ("Antarctic Ozone" 2002). By 2002, the ozone "hole" had grown as large as continental North America. Scientists warned, however, that the rapid shrinkage of areas affected by ozone depletion should not be taken as a pattern. Eskes said that a possibility that one of the two remnants could strengthen and form an expanded area of severe ozone depletion "cannot be excluded" ("Antarctic Ozone" 2002).

In Punta Arenas, a city of 125,000 in southernmost Chile, a "solar stoplight" with four warning levels is used to warn people of ultraviolet levels. Children are taught ozone safety in schools. People wear 50-proof sunblock even on cloudy days. When the solar stoplight is set at orange, the second-highest of four levels, people are told not to expose themselves to the sun more than twenty-one minutes a day between noon and 3 P.M. When the solar stoplight is set at red, most people stay indoors.

"It's a new way of living," said Lidia Amarales Osorno, the Chilean Health Ministry's regional director here. "You'll see the solar stoplight posted in supermarkets, offices and schools, and we even have an Ozone Brigade to raise consciousness about this problem" (Rohter 2002, A-4).

## REFERENCES

Anderson, J. G., W. H. Brune, and M. H. Proffitt. 1989. "Ozone Destruction by Chlorine Radicals within the Antarctic Vortex: The Spatial and Temporal Evolution of $ClO/O_3$, Anticorrelation Based on in Situ ER-2 Data." *Journal of Geophysical Research* 94: 11465–79.

"Antarctic Ozone Hole Shrinks, Divides in Two." 2002. Environment News Service, September 30. http://ens-news.com/ens/sep2002/2002-09-30-03.asp.

"Atmospheric Science: Really High Clouds." 2001. *Science* 292 (April 13): 171.

Austin, J., N. Butchart, and K. P. Shine. 1992. "Possibility of an Arctic Ozone Hole in a Doubled-$CO_2$ Climate." *Nature* 360 (November 19): 221–25.

Baker, Linda. 2000. "The Hole in the Sky: Think the Ozone Layer Is Yesterday's Issue? Think Again." *E: The Environmental Magazine*, November/December, 34–39. http://www.e-magazine.com/november-december 2000/1100feat2.html.

"Biomass Burning Boosts Stratospheric Moisture." 2002. Environment News Service, February 20. http://ens-news.com/ens/feb2002/2002L-02-20-09.html.

Borenstein, Seth. 2000. "Arctic Lost 60 percent of Ozone Layer; Global Warming Suspected." Knight-Ridder News Service, April 6. (In LEXIS)

Brasseur, Guy P., Anne K. Smith, Rashid Khosravi, Theresa Huang, Stacy Walters, Simon Chabrillat, and Gaston Kockarts. 2000. "Natural and Human-Induced Perturbations in the Middle Atmosphere: A Short Tutorial." In *Atmospheric Science across the Stratopause*, edited by David E. Siskind, Stephen D. Eckermann, and Michael E. Summers. Washington, D.C.: American Geophysical Union.

British Broadcasting Corporation. 2000. "Severe Loss to Arctic Ozone." BBC News, April 5. http://news.bbc.co.uk/hi/english/sci/tech/newsid _702000/702388.stm.

Calamai, Peter. 2002. "Alert over Shrinking Ozone Layer." *Toronto Star*, March 18.

Christie, Maureen. 2001. *The Ozone Layer: A Philosophy of Science Perspective*. Cambridge, U.K.: Cambridge University Press.

Connor, Steve. 2000. "Ozone Layer over Northern Hemisphere Is Being Destroyed at 'Unprecedented Rate.'" *London Independent,* March 5.

Crutzen, Paul J. 2001. "The Antarctic Ozone Hole, a Human-Caused Chemical Instability in the Stratosphere: What Should We Learn from It?" 1–11 in *Geosphere–Biosphere Interactions and Climate,* edited by Lennart O. Bengtsson and Claus U. Hammer. Cambridge, U.K.: Cambridge University Press.

Erickson, Jim. 2001. "Boulder Team Sees Obstacle to Saving Ozone Layer; 'Rocks' in Arctic Clouds Hold Harmful Chemicals." *Rocky Mountain News* (Denver), February 9.

Fahey, D. W., R. S. Gao, K. S. Carslaw, J. Kettleborough, P. J. Popp, M. J. Northway, J. C. Holecek, S. C. Ciciora, R. J. McLaughlin, T. L. Thompson, R. H. Winkler, D. G. Baumgardner, B. Gandrud, P. O. Wennberg, S. Dhaniyala, K. McKinney, T. Peter, R. J. Salawitch, T. P. Bui, J. W. Elkins, C. R. Webster, E. L. Atlas, H. Jost, J. C. Wilson, R. L. Herman, A. Kleinböhl, and M. von König. 2001. "The Detection of Large $HNO_3$-Containing Particles in the Winter Arctic Stratosphere." *Science* 291 (February 9): 1026–31.

Farman, J. C., B. G. Gardiner, and J. D. Shanklin. 1985. "Large Losses of Total Ozone Reveal Seasonal ClOx/NOx Interaction." *Nature* 315: 207–10.

Foster, Krishna L., Robert A. Plastridge, Jan W. Bottenheim, Paul B. Shepson, Barbara J. Finlayson-Pitts, and Chester W. Spicer. 2001. "The Role of $Br_2$ and BrCl in Surface Ozone Destruction at Polar Sunrise." *Science* 291 (January 19): 471–74.

Freeman, James. 2001. "Ozone Repair Could Bring New Problem." *Glasgow* (Scotland) *Herald,* April 25.

Gardner, Chester, George C. Papen, Xinzhao Chu, and Weilin Pan. 2001. "First Lidar Observations of Middle Atmosphere Temperatures, Fe Densities, and Polar Mesospheric Clouds over the North and South Poles." *Geophysical Research Letters* 28, no. 7: 1199–1203.

Hartmann, Dennis L., John M. Wallace, Varavut Limpasuvan, David W. J. Thompson, and James R. Holton. 2000. "Can Ozone Depletion and Global Warming Interact to Produce Rapid Climate Change?" *Proceedings of the National Academy of Sciences* 97, no. 4 (February 15): 1412–17.

Henderson, Mark. 2000. "Ozone Hole Will Heal in 50 Years, Say Scientists." *Times* (London), December 4 (in LEXIS).

Jucks, K. W., and R. J. Salawitch. 2000. "Future Changes in Atmospheric Ozone." In *Atmospheric Science across the Stratopause,* edited by David E. Siskind, Stephen D. Eckermann, and Michael E. Summers. Washington, D.C.: American Geophysical Union.

Kerr, Richard A. 1998. "Deep Chill Triggers Record Ozone Hole." *Science* 282 (October 16): 391.

————. 2001. "Stratospheric 'Rocks' May Bode Ill for Ozone." *Science* 291 (February 9): 962–63.

Kirby, Alex. 2000. "Costing the Earth." British Broadcasting Corporation, Radio Four, October 26. http://news.bbc.co.uk/hi/english/sci/tech/newsid_990000/990391.stm.

McFarling, Usha Lee. 2000. "Scientists Warn of Losses in Ozone Layer over Arctic." *Los Angeles Times,* May 27.

————. 2001. "Fear Growing over a Sharp Climate Shift." *Los Angeles Times,* July 13.

Newman, Paul A. 1998. "Preserving Earth's Stratosphere." *Mechanical Engineering,* October. http://www.memagazine.org/backissues/october 98/features/stratos/stratos.html.

Nuttall, Nick. 2000. "Global Warming Boosts el Niño." *Times* (London), October 26 (in LEXIS).

Polakovic, Gary. 2001. "Earth Losing Air-Cleansing Ability, Study Says; Worldwide Decline in a Molecule That Fights Pollution Is Found, but Experts Call the Losses Slight and Not Alarming." *Los Angeles Times,* May 4.

Revkin, Andrew C. 2001. "Study Finds a Decline in Natural Air Cleanser." *New York Times,* May 4 (in LEXIS).

Robock, Alan. 2002. "Pinatubo Eruption: The Climatic Aftermath." *Science* 295 (February 15): 1242–44.

Rohter, Larry. 2002. "Punta Arenas Journal: In an Upside-Down World, Sunshine Is Shunned." *New York Times,* December 27, A-4.

Rowland, Sherwood, and Mario Molina. 1974. "Stratospheric Sink for Chlorofluoromethanes: Chlorine Atom–Catalyzed Destruction of Ozone." *Nature* 249 (June 28): 810–12.

Schrope, Mark. 2000. "Successes in Fight to Save Ozone Layer Could Close Holes by 2050." *Nature* 408 (December 7): 627.

"Scientists Report Large Ozone Loss." 2000. *USA Today,* April 6.

Sherwood, Steven. 2002. "A Microphysical Connection among Biomass Burning, Cumulus Clouds, and Stratospheric Moisture." *Science* 295 (February 15): 1272–75.

Shindell, Drew T., David Rind, and Patrick Lonergan. 1998. "Increased Polar Stratospheric Ozone Losses and Delayed Eventual Recovery Owing to Increasing Greenhouse-Gas Concentrations." *Nature* 392 (April 9): 589–92.

Siskind, David E., Stephen D. Eckermann, and Michael E. Summers, eds. 2000. *Atmospheric Science across the Stratopause.* Washington, D.C.: American Geophysical Union.

Stein, Matthew. 2000. *When Technology Fails: A Manual for Self-Reliance and Planetary Survival.* Santa Fe, N. Mex.: Clear Light.

Stevens, William K. 1999. *The Change in the Weather: People, Weather, and the Science of Climate.* New York: Delacorte Press.

————. 2000. "New Survey Shows Growing Loss of Arctic Atmosphere's Ozone." *New York Times,* April 6, A-19.

Tabazadeh, A., E. J. Jensen, O. B. Toon, K. Drdla, and M. R. Schoeberl. 2001. "Role of the Stratospheric Polar Freezing Belt in Denitrification." *Science* 292 (March 30): 2591–94.

Tabazadeh, A., K. Drdla, M. R. Schoeberl, P. Hamill, and O. B. Toon. 2002. "Arctic 'Ozone Hole' in a Cold Volcanic Stratosphere." *Proceedings of the National Academy of Sciences* 99, no. 5 (March 5): 2609–12.

Thompson, Anne M., Jacquelyn C. Witte, Robert D. Hudson, Hua Guo, Jay R. Herman, and Masatomo Fujiwara. 2001. "Tropical Tropospheric Ozone and Biomass Burning." *Science* 291 (March 16): 2128–32.

Tolbert, Margaret A., and Owen B. Toon. 2001. "Solving the P[olar] S[tratospheric] C[loud] Mystery." *Science* 292 (April 6): 61–63.

Urquhart, Frank. 2001. "Blue Whale Close to Extinction." *Scotsman,* July 19.

Van Emden, Helmut F., and David B. Peakall. 1996. *Beyond Silent Spring: Integrated Pest Management and Chemical Safety.* London: Chapman and Hall and United Nations Educational Programme.

"Volcanic Eruptions Could Damage Ozone Layer." 2002. Environment News Service, March 5. http://ens-news.com/ens/mar2002/2002L-03-05-09.html.

Wang, Chien, and Ronald G. Prinn. 1998. "Impact of Emissions, Chemistry, and Climate on Atmospheric Carbon Monoxide: 100-Year Predictions from a Global Chemistry-Climate Model." Massachusetts Institute of Technology. Joint Program on the Science and Policy of Global Change. Report #35. April. http://web.mit.edu/globalchange/www/rpt35.html.

# 4

# The Chemical Industry, Nonwhite Communities, and the Third World

The dangerous legacy of hazardous synthetic chemical wastes, contaminated manufacturing sites, and polluting industries has fallen disproportionately on poor, nonwhite communities in the United States and burdened them with neighboring hazardous waste sites, incinerators, petrochemical plants, lead contamination, dirty air, and contaminated drinking water.

The burden also has fallen on Third World countries that have become havens for "gift waste," stockpiles of dangerous, largely useless toxic contaminants. This chapter comprises a number of detailed local accounts, with extended reporting on malathion spraying on the Rosebud Reservation in South Dakota to the toxic landscape of Akwesasne, a Mohawk reservation that straddles the border between the United States and Canada, and "Cancer Alley," Louisiana, with its ranks of petrochemical plants.

A 1983 U.S. General Accounting Office study revealed that 75 percent of off-site, commercial hazardous waste landfills in the southeast United States are located within predominately African-American communities. A 1987 study, "Toxic Waste and Race" by the United Church of Christ Commission for Racial Justice (the first national study to correlate waste facilities and demographic characteristics), found that race was the most significant factor in determining where waste facilities are located.

The study said that roughly 60 percent of African-Americans and Latino Americans live in communities with uncontrolled toxic waste sites, and that 15 million African-Americans live in communities with at least one site. For example, the U.S. Army's Pine Bluff Arsenal, which holds 12 percent of the U.S. chemical-weapons stockpile, is located just outside Pine Bluff, Arkansas, which is 53 percent black. A proposed $200 million incinerator under construction in Pine Bluff would be the nation's second chemical-weapons incinerator. Nearly 30 percent of Pine Bluff's residents earn less than a poverty-level wage as defined by the 1990 U.S. Census.

A 1992 study by the *National Law Journal,* "Unequal Protection," uncovered significant disparities in the way the U.S. Environmental Protection Agency (EPA) enforces laws (*National Law* 1992). The study found abundant evidence of a racial divide when the U.S. government cleans up toxic waste sites and punishes polluters. White communities see faster action, better results, and stiffer penalties than communities where blacks, Latinos, and other minorities make up most of the population. The handful of descriptions that follow are only a small sampling of the hot spot situations in minority communities.

## MALATHION AND THE ROSEBUD SIOUX IN MISSION, SOUTH DAKOTA

The town of Mission, on the Rosebud Sioux reservation in South Dakota, was sprayed routinely with malathion until at least the middle 1990s. In 1995, the man who sprayed the town died of cancer, according to Joe Allen, editor of the *Circle,* a Native American newspaper published in Minneapolis (Allen 1995).

The *Circle* published detailed accounts of Mission's spraying and the miseries it caused Native American residents there. The article began by comparing the massive press coverage provoked in the United States by the nerve gas sarin in a Tokyo subway with widespread ignorance that a derivative of the same chemical, malathion, was being sprayed liberally on a South Dakota Indian reservation.

People living at Rosebud described the peculiar hissing sound of the spraying machine as it was driven on its rounds, spreading a dirty gray fog along the reservation's dirt roads. When local residents confronted the sprayers, they were told the reservation was being treated for mosquitoes. Soon, everything in their homes smelled of the chemical—clothes, unwrapped food, furniture, curtains, and more.

One Rosebud resident, Jane Kirby, who was pregnant at the time, was stricken with splitting headaches after spraying of her neighborhood. She took no medication because of her pregnancy, but her husband, who also had headaches, took several doses of aspirin. They were no help. The same night, the Kirbys' daughter Gemma, two years old, awoke at night breathing in raspy gasps.

The next morning, the Kirbys called Mission's mayor, Jack Herman, and demanded an end to the spraying. Herman refused and laughed in their faces. A physician's assistant provided the Kirbys with information on malathion; they were surprised to learn that it had been developed by Nazi Germany as a possible nerve-gas weapon. Four days after Mission's mayor had laughed in her face, Jane went into labor, as her husband became progressively dizzier, weaker, and prone to muscle cramps. Within a few days, the Kirbys learned that fourteen other people on their street were suffering similar symptoms.

On July 9, 1992, a city crew appeared at the Kirbys' door with a notice that spraying would resume the next day. The spray crew's members pledged to avoid the couple and their one-day-old infant. The Kirbys had asked that their area be spared the spraying because of the infant, but Mayor Herman again refused. He said the machine spraying the malathion could not be shut down easily for one home on a given street. Instead of enduring the spraying, the Kirbys spent the next night at a friend's mouse-infested cabin in the nearby countryside. Back in town, all during the summer spraying season, people complained of headaches and of waking in the night gasping for breath.

During summer 1992, people at Rosebud also began to trade accounts of illnesses associated with pesticide spraying, including an account of a man in his twenties and his dog who lived in the tiny community of White Horse, near Mission. Both were found dead in their home after the house was sprayed for cockroaches by an exterminator hired through the reservation housing authority.

During summer 1993, Anna Carol Thin Elk, fifty-one, of Mission, was caught in a malathion spraying. She thought little of it and wore the same clothes the next day without having taken a shower. By noon that day, pain was spreading through Thin Elk's wrists and she experienced waves of chills. Severe pain spread through her entire body during the next three days, until she was wracked by fever and leg spasms. She also experienced acute difficulty breathing. Thin Elk was taken to Rosebud Hospital, where she noticed, while waiting, that three or four children in the room also were having trouble breathing. The first question the doctor asked each of their mothers was, Have you been near the area being sprayed for mosquitoes? By the time a doctor saw Thin Elk an hour and a half later, she was semiconscious and nearly asphyxiated. Thin Elk was then taken by ambulance to Rapid City General Hospital, three hours away, the closest medical facility possessing the proper antidote for malathion poisoning.

Thin Elk was discharged from the hospital the next day. She had no medical insurance, no spare clothes, and no way to get back to Mission other than a ride that was offered to her that day. For a week, Thin Elk struggled to keep working with severe fatigue and dizziness, as her employers denied her sick leave. When Thin Elk complained to Mayor Herman about how the pesticides affected her, according to the *Circle*'s account, "I was told I should move out of town if I didn't like it" (Allen 1995, 10).

The *Circle* published several other, similar accounts of Mission residents who became ill from the spraying, some of whom experienced periods of near paralysis. In the meantime, Ed Einspar, who had sprayed most of the malathion on Mission, died. One day, shortly before he died, Einspar completed his spraying rounds without wearing a gas mask. Shortly thereafter, he suffered a severe gastrointestinal illness that aggravated his chronic asthma and emphysema. Einspar's niece, six-year-old Fianna White Hawk, suffered a type of chronic pneumonia that has been attributed to malathion exposure (Allen 1995).

Malathion was being sprayed in Mission despite the fact that the chemical itself posed a much larger human-health risk to people at Rosebud than to the mosquitoes it was being used to eradicate. The area has no record of mosquito-borne disease, and so the insects are more an irritant than a serious threat to human or animals' health. Some Mission town officials were quoted in the *Circle* article saying that AIDS could be spread by mosquitoes. No one knows who told them that, or whether they believed it. The City of Minneapolis decided in 1982 that malathion was little good against mosquitoes, because its effects on them largely vanished with the next substantial rainfall and egg-laying cycle. Salesmen of malathion-contaminated sprays were using mosquito-phobia to sell a dangerous product.

The malathion sprayed at Mission probably was given to the town by the U.S. Department of Agriculture, which has, at various times, had a policy of forwarding surplus pesticides to municipal governments. Malathion is especially dangerous to alcoholics, because the liver of most alcoholics has been damaged beyond its ability to deal with the pesticide, which then overwhelms the body. It is also nearly as dangerous inhaled as absorbed through the skin. Roughly 75 percent of Rosebud's population suffers from alcoholism, according to Mission's Little Hoop Lodge Treatment Center (Allen 1995).

## AKWESASNE: LAND OF THE TOXIC TURTLES

For several centuries of human occupancy, the site the Mohawks call Akwesasne was a natural wonderland: well watered; thickly forested with white pine, oak, hickory, and ash; home to deer, elk, and other game animals. The rich soil in the bottomlands of a valley into which several rivers flowed allowed farming to flourish. The very name that the Akwesasne Mohawks gave their territory about 1755 testifies to the bounty of the land. Akwesasne in the Mohawk language means "Land Where the Partridge Drums," after the distinct sound that a male ruffed grouse makes during its courtship rituals. Lying at the confluence of the Saint Lawrence, Saint Regis, Racquette, Grass, and Salmon rivers, Akwesasne, until recent times, also provided its human occupants with large runs of sturgeon, bass, and walleye pike.

Within roughly half a century, this land of natural wonders has become a place where one cannot eat local fish and game, because their flesh now is contaminated at toxic levels with anthropogenic carcinogens such as PCBs. In some places, one cannot drink the water, for the same reason. In parts of Akwesasne, residents have been told to plow under their gardens and to have mothers' breast milk tested for contamination. In place of sustaining rivers and a land to which the Mohawks still offer thanksgiving prayers, late-twentieth-century capitalism has offered incinerators and dumps for medical and industrial waste. Akwesasne, which straddles New York State's border with Quebec and Ontario, has become the most polluted native reserve in Canada and a number-one toxic site in the U.S. EPA's Superfund list of sites badly needing cleanup.

Within the living memory of many people at Akwesasne, the Land Where the Partridge Drums, has inherited the toxicological consequences of General Motors (GM) waste lagoons in which animals have been found with levels of PCBs in their fat that qualifies them as toxic waste under U.S. EPA guidelines. Akwesasne has become riskier to human health than most urban areas, a place where any grouse still living may be more concerned about its heartbeat than about its drumbeat (Johansen 1993).

Soon after the St. Lawrence Seaway opened during the middle 1950s, GM, Reynolds Metals Company, and the Aluminum Company of America (Alcoa) built plants directly upstream of Akwesasne. According to the state attorney general's office, GM never obtained a permit to operate its dumpsites (Thomas 2001).

Paul Thompson of Akwesasne remembers his childhood in a more innocent time: In a cove off the St. Lawrence River, walleye pike leaped upstream to spawn every April. His family bought fresh catch from fishermen on the river's banks, as they "peer[ed] into their crates and pick[ed] out the evening supper: a perch, bass, or maybe a sturgeon head for soup" (Sengupta 2001). Nearby sits a mound in which Thompson's brothers and sisters once had foraged. "They plucked scrap metal and sold it in town for extra cash. They burned the wood at home" (Sengupta 2001). At the time, no one at Akwesasne realized that the nearby GM engine-parts factory, built during the 1950s, was turning the fish to toxic waste and the children's play mound into a toxic dump.

Nearby, Turtle Cove, an inlet leading into the St. Lawrence River, was a favorite swimming hole for children at Akwesasne. In the spring, boys, like generations before them, learned to spear bullhead pike making their way through the cove to spawn. The cove, which is a few feet from the GM foundry, is a swimming hole no longer. Instead, it is one of GM's toxic-waste dumps.

Dana Leigh Thompson grew up with a forty-foot GM waste heap as a neighbor. The toxic hill slopes into Containment Cove, a local swimming hole until tests revealed PCB levels many times toxic limits. "There were three big rocks out there," Thompson said. "When we taught kids how to swim, they could swim out to the middle and stand. It was an achievement" (Seely 2001).

Thompson and other local residents began to suspect toxicity in GM's waste dumps during the middle 1970s, but GM continued to dump PCBs in the area without a state permit until 1986. Cleanup efforts began about 1988 but have stalled over differing approaches to the problem.

The installation of the temporary cap, during 1983, initiated Thompson's former playground into the ranks of federal Superfund sites, as one of the most toxic (in this case, PCB-contaminated) patches of ground in North America. During 1988, according to residents of Akwesasne, "A crew of men, covered head to toe in white spaceman-like suits, covered it [the mound] with an impermeable sheath" (Sengupta 2001). Meant to be temporary (in place until the dump was cleaned up) the capped mound remained in place thirteen years later.

At first, "I didn't even know what PCBs were," said Jim Ransom, an Akwesasne resident and director of the Haudenosaunee Environmental Task Force, an environmental group that advocates on behalf of all Iroquois. "There was a high level of concern, but I think that there was also a lot of unknowns because people didn't know what this chemical was and what it could do to us." By the mid-1980s, preliminary testing showed that it was no longer safe to eat fish and wildlife caught in some areas of the reservation. Sheree Bonaparte, then a young mother with a farm near the GM landfill, laughed when GM first distributed bottled water to residents (Thomas 2001). At first, she said, "Everybody kind of thought it was ridiculous," she said. "The water comes from the earth and it seemed silly to go get it from a bottle" (Thomas 2001).

The Mohawks, state agencies, and area universities soon began studying PCB levels in breast milk and in infants. "Those studies proved beyond any shadow of a doubt that at the beginning of the study, the Mohawks had significantly higher levels of PCBs," said David Carpenter, a professor of environmental health and toxicology at the State University of New York at Albany (Thomas 2001). Carpenter said that the Akwesasne Mohawks have higher-than-average rates of some diseases that are associated with PCB contamination. One such disease, hypothyroidism, is "strikingly elevated," Carpenter said (Thomas 2001). PCBs disrupt production of thyroid hormones, which leads to hypothyroidism, a disorder that can cause mental dullness, obesity, and learning disabilities in children.

For more than a decade, GM and the Akwesasne Mohawks have debated how best to clean up the company's waste lagoons. The company has suggested sealing the dumps permanently in place, meanwhile also building a wall to prevent existing PCBs from migrating to other parts of Akwesasne. Federal officials have approved this plan, but GM requires access to the reservation to build the wall. The Mohawks have denied access because they believe GM is seeking a relatively inexpensive way out of a problem that requires removal, completely, of all soil tainted by PCBs dumped there.

In the meantime, the U.S. EPA commended GM for moving diligently to clean up its waste sites. GM found itself inching toward agreement with the Mohawks' solution. The company, for example, did remove 23,000 cubic yards of polluted sediment from the St. Lawrence River during 1995. During 2000 GM excavated contaminated sludge from inactive lagoons. By the end of 2001, the company was planning to have removed soil from the banks of the Racquette River.

"This is the only place we have, and we're going to be here forever," explained Ken Jock, director of the St. Regis Mohawk tribe's environmental division. "Our teachers have told us, when we make a decision we have to look at how it affects the next seven generations. It's a different sense of time" (Sengupta 2001). Before the area was so widely contaminated, fishing, hunting, and trapping "were something our parents had pride in handing down to our children," Jock said. Because Akwesasne residents have been advised by state officials not to eat local fish or game, fishing and hunting skills are being lost. "It's pretty important to our identity as a people" (Thomas 2001).

Scientists have concluded that even low levels of PCB exposure here could cause more serious illnesses than previously thought. "That small relationship we expect to see correlated with reduced I.Q., with poor performance in school, with some abnormality in growth, particularly sexual maturation, and increased susceptibility to certain chronic diseases such as thyroid disease and diabetes," said Carpenter. "This has adversely affected their health" (Sengupta 2001).

Thompson's family has been wracked by illnesses that once were very rare at Akwesasne. Thompson himself has diabetes. Four of his five siblings have thyroid disorders of a type often aggravated by PCBs. Thompson's sister Marilyn had her thyroid gland removed when a tumor was discovered there. All six of her children have asthma; two of them also have learning disabilities; another suffers from a thyroid condition. A two-year-old granddaughter of Thompson was born with a muscle disorder that has affected her motor skills. Another family member has experienced fourteen miscarriages (Sengupta 2001).

Rowena General of Akwesasne said the contamination has led to a "health crisis" for the more than 10,000 people who live on the reservation. "Recent analysis of clinical and hospital records on the reservation shows an epidemic of thyroid problems and also the incidence of cancers, diabetes and respiratory diseases is higher than average," General said (Associated Press 2001).

The distribution of certain chronic diseases at Akwesasne was determined using computerized medical records of the St. Regis Mohawk Health Services Clinic. Prevalence proportions, annual incidence rates, and five-year incidence rates were computed for the period January 1, 1992, to January 1, 1997, for asthma, diabetes mellitus type II, hypothyroidism, and osteoarthritis. The study indicated that hypothyroidism and diabetes

showed higher age-specific prevalence than in the general U.S. population. Osteoarthritis was extremely frequent among people 60 years of age and older, and it may also be elevated in prevalence in relation to the U.S. general population. The incidence and prevalence trends of diabetes type II and osteoarthritis were stationary, but those for asthma and hypothyroidism showed increases over the study period. Morbidity from asthma and acquired hypothyroidism should be monitored in the future and investigated through analytic epidemiologic methods for a possible association with lifestyle and environmental factors. (Negoita et al. 2001, 84)

A study conducted at Cornell University indicated that smokestack effluvia from the Massena Reynolds Metals factory also destroyed once-profitable cattle and dairy farms in Cornwall on the Ontario side of Akwesasne. The study linked fluorides to the demise of cattle as early as 1978. Many of the cattle, as well as fish, suffered from fluoride poisoning that weakened their bones and decayed their teeth. Ernest Benedict's Herefords died while giving birth, while Noah Point's cattle lost their teeth and Mohawk fishermen landed perch and bass with deformed spines and large ulcers on their skins. The fluoride was a by-product of a large aluminum smelter in Massena, New York, that routinely fills the air with yellowish gray fumes smelling of acid and metal (Krook and Maylin 1979).

Although Reynolds Metals, owner of the aluminum smelter, cut its fluoride emissions from 300 pounds an hour in 1959 to 75 pounds an hour in 1980, the few cattle still feeding in the area continued to die of fluoride poisoning. The pollution of Akwesasne is accentuated by the fact that most of the plants emitting toxins are west of the reservation, upstream and often upwind.

Chronic fluoride poisoning in Cornwall Island cattle was "manifested clinically by stunted growth and dental fluorosis to a degree of severe interference with drinking and mastication. Cows died at or were slaughtered after the third pregnancy. . . . Concentrations exceeding 10,000 p.p.m. fluoride were recorded in cancellous bone of a 4- to 5-year-old cow" (Krook and Maylin 1979, 1).

During late March 2001 New York Attorney General Eliot Spitzer and the St. Regis Mohawk Nation gave notice to GM that he would sue the company in federal court unless it began cleaning up two PCB dumpsites at its Massena plant within ninety days. In a letter to GM, Spitzer said if the company does not make substantial progress within ninety days, he would ask a U.S. District Court judge in Albany to declare the site an "imminent and substantial endangerment" and order an immediate cleanup (Associated Press 2001). "General Motors has been on notice since at least 1980 that PCBs were being released into the St. Lawrence River and onto the St. Regis Mohawk Reservation from its two hazardous waste dumps," Spitzer said. "The company also has known for the past 15 years that the landfills may endanger public health and the environment. Despite this knowledge, GM has failed to control the release of these toxins from its property" (Associated Press 2001).

By June, barely within its ninety-day deadline, the attorney general's office said that GM was taking PCB-removal talks with a new sense of seriousness. "If we have to, we are ready to file a lawsuit at a moment's notice, and G.M. knows that we are prepared to do so, if necessary," said Marc Violette of the New York Attorney General's office ("State, G.M." 2001).

Chris Amato, the assistant New York attorney general working on the case, said the injustice is clear. "This is another example of a Native American community being treated as second-class citizens," he said. "I guarantee you if this site was located next to a very middle-class, white neighborhood, this site would be well on its way to being remediated" (Thomas 2001).

"General Motors' illegal industrial waste dump has been poisoning the Mohawk people for over 50 years," said Akwesasne Mohawk Loran Thompson. "Despite all of our efforts, the G.M. facility continues to discharge toxic contaminants into the Akwesasne environment. General Motors is guilty of environmental injustice and they have been completely negligent in overlooking the damages to the health, well-being, economy, and lifestyle of the Mohawk people" (Associated Press 2001).

## OUR DAILY DIOXIN: LIFE AND DEATH IN LOUISIANA'S "CANCER ALLEY"

The nickname Cancer Alley has been applied to a stretch of the Mississippi River between Baton Rouge and New Orleans, Louisiana, which includes several mostly low-income, mainly African-American communities in the shadows of petrochemical and plastic industries. Fourteen of fifteen U.S. plants that produce vinyl chloride monomer (VCM) and ethylene dichloride (EDC), the basic building blocks used to make polyvinyl chloride (PVC), are in Louisiana and Texas. A large number of the incinerators that burn discarded PVC products also are located in low-income minority communities. Dioxin and dioxin-like pollutants, including PCBs, are unintentionally produced and released into the air and water and in hazardous waste during the manufacture of VCM, the incineration of vinyl products, and the burning of PVC in accidental fires.

The manufacture of polyvinyl chloride produces copious amounts of dioxins, which "have been linked to immune-system suppression, reproductive disorders, a variety of cancers, and endometriosis," according to one observer, who continued: "Dioxins are an unavoidable consequence of making PVC. Dioxins created by PVC production are released by on-site incinerators, flares, boilers, wastewater treatment systems and even in trace quantities in vinyl resins" (Costner 1995). Products manufactured with PVCs "create dioxins when burned, leach toxic additives during use . . . and are the least recyclable of all major plastics" (Cray and Hardin 1998). "The production of the carcinogenic monomer is what results in the highest levels of dioxin release," says Charlie Cray, a Greenpeace toxics campaigner (Shintech 1999).

As early as November 1983, a U.S. EPA contractor (Versar 1983) learned that the production of vinyl feed stocks (ethylene dichloride and vinyl chloride monomer—EDC/VCM) was an inadvertent source of PCB production. Hinting that the entire lifecycle of PVC should be examined more intensely for PCB contamination, Versar also concluded that "it would be necessary to consider input PCBs contaminating chlorinated feed stocks in further downstream processing" (Dioxin Deception 2001). In 1990, however, at the request of the Vinyl Institute and other VCM producers, the EPA deleted dioxin from the list of constituents of concern in a VCM waste stream because of "the costs of analysis and the reluctance of waste-treatment facilities to take wastes designated as dioxin-contaminated" (Dioxin Deception 2001).

During 1987, 106 residents of Reveilletown, Louisiana, a small African-American community about ten miles south of Baton Rouge, filed a lawsuit against Georgia Pacific and Georgia Gulf arguing that they had suffered health problems and property damage. After settling out of court for an undisclosed amount, Georgia Gulf relocated the remaining families and then tore down every structure in town. Management at Dow Chemical's neighboring factory in Plaquemine followed suit soon afterward, buying out all the residents of the small town of Morrisonville (Bowermaster 1993).

The PVC plants incinerate dioxin-contaminated wastes, and in so doing, according to Greenpeace, a portion of the originally discharged dioxins are emitted undestroyed, and new dioxins are created as by-products of the incineration process. The presence of copper and other metals in PVC industry wastes can act as a catalyst to further increase dioxin formation (Duchin 1997). Under usual operating conditions, several chemicals are burned in a constantly changing mixture, creating a variety of synergistic chemical-thermal reactions and emissions.

Greenpeace has called on the U.S. EPA to "impose a moratorium on permits for new vinyl facilities or expansion of existing facilities, and to modify permits at existing plants to require that dioxin releases to all media, including waste destined for disposal, be brought to zero within five years" (Duchin 1997). Greenpeace advocates a ban on the use of PVC in many types of toys, as well as furniture, wallpaper, and medical devices such as intravenous bags. Moreover, Greenpeace has called for a ban of PVC in products that may be susceptible to fire, such as cabling and other construction materials, notably in appliances and vehicles; and "metals with PVC residues that are recycled in combustion-based processes (i.e., automobiles)" (Duchin 1997).

During September 1998 Greenpeace activists joined the people of Covenant, Louisiana, three-quarters of whom are black, in celebration of the news that the Japanese chemical giant Shintech would not, after all, build an enormous, 3,000-acre polyvinyl chloride (PVC) factory in their town. Shintech, a Japanese chemical company, had been trying since 1996 to locate three factories and an incinerator near homes and schools in Covenant. Local people argued against location of the plant in their town on civil-rights grounds; they argued that Louisiana authorities would violate federal civil-rights laws if they licensed the Shintech plant in a predominantly African-American community where pollution is already making people sick (Cray and Harden 1998). This struggle was undertaken with the idea of providing other poor minority communities with legal precedents that would be useful in fending off expansion of environmentally intrusive industries.

The proposed PVC plant would have been one of the largest of its type in the world. Local residents worried most about anticipated emissions of polyvinyl chloride (PVC), ethylene dichloride (EDC), and vinyl chloride monomer (VCM) which "may reasonably be anticipated to result in an increase in mortality or an increase in serious irreversible, or incapacitating reversible illness. Vinyl chloride is a known human carcinogen which causes a rare cancer of the liver" (U.S. Environmental Protection Agency 1998). According to the company, the proposed Shintech manufacturing plant would have released about 600,000 pounds of toxic chemicals into the air per year and would have poured nearly 8 million gallons of toxic wastewater each day into the Mississippi River, which provides drinking water for the city of New Orleans (Public Notice n.d.).

Greenpeace's campaign against dioxins began during the 1980s. In 1988 the Greenpeace ship MC *Beluga* toured Cancer Alley. In 1989, Greenpeace also released the report *We all Live Downstream,* documenting serious chemical pollution in the Mississippi River (Costner and Thornton 1989). During the middle and late 1990s, Greenpeace activists, often swimming at night in polluted water, sampled effluent from several U.S. vinyl producers and found dioxins at each of the twenty-seven sites tested.

One sample, obtained at the Vulcan Chemicals facility in Geismar, Louisiana, contained 6 p.p.m. of TEQ dioxin, a level as high as the historic levels in wastes left from Agent Orange production (Dioxin Deception 2001). "These data are important, since they add to the growing body of evidence pointing to the lifecycle of polyvinyl chloride (PVC) plastic as one of the largest single sources of the nation's total dioxin burden" (Duchin 1997).

Concentrations of dioxins in some of the samples taken by Greenpeace were extraordinarily high:

• Vulcan Chemicals, Geismar, Louisiana: 200,750 p.p.b. dioxins in a sample of "heavy end" waste

- Formosa Plastics, Point Comfort, Texas: 761 p.p.b. dioxins in a sample of "heavy end" waste

- Georgia Gulf, Plaquemine, Louisiana: 1,248 p.p.b. dioxins in a waste sample from a tank labeled to contain "heavy ends," "tars," and other similar types of highly contaminated wastes (Duchin 1997)

On occasion, Greenpeace also blockaded trains transporting vinyl chloride at PPG Industries with two buses as it called for global elimination of dioxin and other persistent poisons. Of twenty-eight residents whose blood was tested by the federal agency, twelve had elevated levels of dioxin or similar substances ("Louisiana Town" n.d.). A Greenpeace Web page asserted that "Cancers, respiratory problems, reproductive disorders are among those illnesses associated with dioxin contamination, and which are occurring with alarming frequency in Mossville" ("U.S. Government" 1999). By 1998 and 1999 Greenpeace was organizing toxic patrols with community activists to monitor chemical plants along the Mississippi and on the shores of Lake Charles.

Residents in Covenant turned their backs on arguments that the plant would provide them more jobs and economic prosperity. They argued that the town already hosts a bevy of high-technology industries that emit various toxins, yet 40 percent of the townspeople live below the poverty line. Most of the plants are highly automated with little need for workers possessing marginal skills. Instead, the plants tend to hire a few people (usually from the outside) with computer skills and a working knowledge of physics and chemistry. Shintech's controller, Dick Mason, said the new complex would bring with it 165 permanent jobs; the company also promised to spend $500,000 for local job training, even though Louisiana officials had agreed to give Shintech $130 million in tax breaks without setting any specific goals to commit the company to local hiring.

The state of Louisiana strongly supported Shintech's plans; its air, water, and coastal zone permits were granted, swiftly, "with the backing of nearly all the relevant state and local officials, and over the objections of over 18 groups and a large segment of the local population" (Sierra Club n.d.). In a precedent-setting decision during September 1997, the U.S. EPA rejected the state air permit. The EPA found forty-nine deficiencies in the state's granting of the permit. The EPA also challenged state officials on questions of environmental racism in siting the plant. Represented by the Tulane Environmental Law Clinic, citizens' groups filed appeals to the air, water, and coastal-zone permits, "citing official bias as well as technical and legal problems" (Sierra Club n.d.).

Local citizens also lodged a complaint under Title VI of the U.S. Civil Rights Act. The thirty-three-page complaint

[A]sserts that statements and actions by Governor Foster, DED Secretary Kevin Reilly (the Governor's liaison on the Shintech matter), Secretary Givens and other DEQ officials and employees create bias, prejudice or interest toward Shintech and against the citizen groups and eliminate the legally-required appearance of complete fairness and impartiality. The pervasive evidence of bias or prejudice includes threats and investigations by the Governor and Mr. Reilly against plant supporters and the Tulane Environmental Law Clinic, and an extensive, taxpayer-financed effort by the DEQ Secretary's office to organize and assist a group in St. James to support Shintech and oppose the efforts of the community groups that object to the proposed facility. (Sierra Club n.d.)

Covenant's situation was grist for precedent; a review of 1990 census data "shows that communities living near the nation's existing 15 EDC/VCM facilities have a 55 percent higher percentage of people of color than the national average, and a 24 percent lower per capita income than the national average" (Sierra Club n.d.).

After the EPA rejected its air permit, the Louisiana Department of Environmental Quality was required, by law, to convene public hearings on Shintech's application. Nearly three hundred citizens attended the hearings in Covenant on December 9, 1996. Testimony continued for more than eight hours, according to one account, "at least 95 percent" against Shintech (Sierra Club n.d.). "Enough is enough," testified Patricia Melancon, president of the local group, St. James Citizens, "This is a low-income area, and less than half of the adults in this community have a high-school diploma. We will get all of the pollution, but none of the jobs" (Cray 1996).

To demonstrate public support at the hearing, Shintech flew in more than forty employees from its Freeport, Texas, plant as well as its paid consultants, whose testimony centered on potential employment and tax revenue from the company's proposal. "Why are they [Shintech] speaking of economic development at an air permit hearing?" asked Kishi Animashaun, Greenpeace Toxics campaigner (Cray 1996).

After withdrawing from Covenant, Shintech switched its building plans to Plaquemine, thirty miles north, where its executives hoped that local opinion would be more accommodating. Shintech said it planned to buy vinyl chloride monomer (feedstock finished PVC) from a Dow Chemical plant in Plaquemine. People in Plaquemine then organized People Reaching Out to Eliminate Shintech's Toxins (PROTEST). In their opposition to new PVC manufacturing there, the members of PROTEST cited already-high cancer rates in the town, falling property values, and a lack of emergency evacuation routes should an accident occur. Liz Avants, speaking for PROTEST, said, "Every day, we hear about more cases of cancer and other health effects" (Shintech 1999).

A study released during 1999 by the Agency for Toxic Substances and Disease Control (ATSDC) indicated that dioxin levels in Mossville, one community in Cancer Alley (near Lake Charles), showed that the average concentration of dioxins and PCBs in the blood of its residents were three times the average level in the general population.

On March 5, 1999, sixty Greenpeace activists from twenty-two nations converged on the Louisiana State House to call attention to global health and environmental threats caused by Louisiana's numerous polluting PVC production facilities. Dressed in T-shirts with the slogan "Love Louisiana but Not PVC" in sixteen languages, the Greenpeace Toxic Patrol called on Governor Mike Foster to clean up the state's Cancer Alley. The group then marched with local environmental activists from the Capitol to the governor's mansion to deliver a "lunch" of contaminated fish and water from some of the state's most polluted waterways ("Louisiana's Cancer" 1999).

Greenpeace on June 22, 1999, launched a Toxics Patrol bus trip to more than a dozen of the state's chemical facilities whose emissions have made Louisiana a global toxic hot spot. The bus tour began with stops at three controversial facilities: Rhodia, the nation's first napalm incinerator; Formosa, which makes components of vinyl; and Dow Chemical, another vinyl producer that planned to join with Shintech on a newly proposed vinyl plant in West Baton Rouge. With representatives of several state and local environmental groups, Greenpeace displayed posted signs at chemical facilities warning that toxic pollution "does not stop at the fence" ("With Public's" 1999).

"Louisiana ranks number one in the nation in per-capita toxic releases to the environment, and her citizens are bearing a terrible health burden for it," said Greenpeace toxics campaigner Damu Smith. "Our Toxics Patrol is out to expose some of the state's worst toxic offenders. . . . Louisiana is at the center of the nation's growing problem of environmental racism and injustice" ("With Public's" 1999).

Acclaimed author Alice Walker, actress Alfre Woodard, actor Mike Farrell, Rev. Al Sharpton, and members of Congress Maxine Waters and John Conyers toured Cancer Alley on June 9, 2001 ("Celebrities" n.d.). After the walking tour, the delegation convened in New Orleans for the first-ever national town meeting on environmental justice with residents, government officials, and representatives of the chemical industry. During the town meeting, the delegation heard personal testimonies of industrial contamination victims who live in Cancer Alley communities that are experiencing illnesses and social problems associated with toxic pollution. The town meeting was coordinated and sponsored by Greenpeace ("Celebrities" 2001).

Bill Moyers's report "Trade Secrets," aired on the Public Broadcasting System on March 26, 2001, described the lives and deaths of some Cancer Alley workers, such as Ray Reynolds, forty-three years old, who was shown in the living room of his house a few miles from the chemical plant where he worked for sixteen years. Reynolds was shown dying of toxic neuropathy that had spread from his nerve cells to his brain. Another worker, Dan Ross, made his living for twenty-three years producing the raw vinyl chloride that is basic to the manufacture of PVC plastic. In 1989 Ross was told he had a rare form of brain cancer.

Corporate documents displayed by Moyers indicated that manufacturers of PVCs knew as early as 1959 that "500 parts per million is going to produce rather appreciable injury when inhaled seven hours a day, five days a week for an extended period" (Moyers 2001). As the years went by, the level at which exposure to PVC and other organochlorines were believed to cause injury to the human body crept downward. In 1959 workers were regularly exposed to at least 500 p.p.m. during their work shifts. Some workers described standing in clouds of the chemical at levels much above 500 p.p.m. Some X-rays showed workers' bones dissolving.

"Trade Secrets" included interviews with people affected by dioxins resulting from the manufacture of vinyl chloride, which draw on an archive of secret and confidential documents unearthed in a lawsuit by the widow of a Louisiana chemical worker. A 1959 memo to the B. F. Goodrich Company, for example, says vinyl chloride "is going to produce rather appreciable injury when inhaled seven hours a day, five days a week for an extended period" (Kurtz 2001, C-1).


## U.S. TOXIC WASTES IN CANADA

Canada has been accepting shipments of PCB-contaminated waste from American military bases in foreign countries that the United States itself will not accept. Greenpeace International released a U.S. Defense Department document prepared for the U.S. Congress during March 1999 that refers to "current shipments of foreign-manufactured waste to Canada" (Rusnell 2000). "Greenpeace will be asking for a detailed summary of all the American military waste that has been coming into Canada and for how long," said Darryl Luscombe, a toxic-waste campaigner in Ottawa who obtained the document. "This also raises the question of where this waste has been going in Canada" (Rusnell 2000).

A U.S. company, Trans-Cycle Industries, was known to have shipped eighty-one tons of PCB-contaminated waste from a U.S. military base in Japan to Canada. The shipment, aboard a Chinese bulk carrier, arrived at the port of Vancouver, British Columbia, on April 8, 1999. The waste was then trucked to the company's plant at Kirkland Lake in northern Ontario. Trans-Cycle had no approval from Ontario to import hazardous waste from outside Canada. In fact, Ontario's Environment Department had denied the company's foreign-waste import application.

In 1997 the United States closed its borders to PCBs. Under Canadian environmental law, waste containing fewer than 50 parts per million of PCBs can be imported from any of the 130 countries that signed the Basel Convention, an international agreement meant to control the movement of hazardous waste around the world.

More than 200 truckloads of U.S. toxic waste undergoing treatment in Alberta became a target of controversy there as environmentalists and a city alderman wondered whether an accident might occur as the waste was en route to Swan Hills, 500 kilometers northwest of Calgary. "It could cause a tremendous health hazard if you had a truckload of [persistent organic chemicals] or PCBs go down," said Margaret Chandler, editor of the Alberta environmental magazine *Encompass* (Rusnell 2000).

Bovar Inc.'s Swan Hills Treatment Center is finalizing contracts and permits. Swan Hills was being prepared to receive POPs such as DDT and other pesticides from the Pacific Northwest. Bovar has been given government approval to treat PCBs and other toxic chemicals from outside Canada. During 2000, Bovar burned 3,000 to 5,000 tons—up to 250 truckloads—of foreign toxic waste at its facility, according to its president, John Kuziak. "We cannot afford to have an incident on the roads. That would really cause people to be concerned, so we take very careful precautions," Kuziak said (Pierson 2000, 4).

## "GIFT WASTE" AND THIRD WORLD COUNTRIES

Tanzania has become home to huge stockpiles of agricultural pesticides and veterinary drugs, including such compounds as DDT, which pose a threat to the environment and to the public's health. Some stocks are more than thirty years old and are stored with few or no safety precautions. Stocks of toxic chemicals have been growing in Tanzania because there is no environmentally sound way to dispose of them. "Environmental and health experts warn that unless quick action is taken, the situation could be catastrophic," according to a *Los Angeles Times* report (Simmons 2000). Environmentalists estimate that Tanzania has more than five hundred tons of agricultural compounds dumped or stored at more than a hundred sites (Simmons 2000). Most of the material was imported more than a decade ago as donations by China, Japan, Italy, the United States, and other wealthier countries seeking to dispose of chemicals that now are illegal, and unusable, at home.

The scope of the gift waste problem has been growing. During early May 2001 the United Nations Food and Agriculture Organization (FAO) said that more than 500,000 tons of banned or expired pesticides are seriously threatening the environment and the health of millions of people worldwide in developing countries. This was five times FAO's previous estimate. "In Asia," according to an FAO news release, "The quantities of obsolete pesticides are estimated at over 200,000 tons, in Africa and the Near East at over 100,000 tons, and in Eastern Europe and the former Soviet Union at more than 200,000 tons" (FAO 2001).

"The lethal legacy of obsolete pesticides is alarming and urgent action is needed to clean up waste dumps," said Alemayehu Wodageneh, an FAO expert on obsolete pesticides. "These 'forgotten' stocks are not only a hazard to people's health but they also contaminate natural resources like water and soil. Leaking pesticides can poison a very large area, making it unfit for crop production" (FAO 2001). In its new report, FAO calls upon chemical companies represented by the Global Crop Protection Federation (GCPF), to aid global disposal of pesticides produced by GCPF member companies. "Support from industry is crucial for the future disposal of pesticides because aid agencies of donor countries cannot cover all the costs without a substantial contribution from industry," the FAO expert said (FAO 2001).

During the mid-1970s to 1980s, pesticides were sold to many African nations to improve agricultural production. At the time, the pesticides were offered "hand in hand with friendly donors wanting to help," when the effects of the toxins were not widely known (Simmons 2000). Today, "much is stockpiled in tattered sacks or in leaking and corroding metal drums near settlements. Some of the stores are located near rivers, irrigation schemes or ports; others are stashed outdoors in mountainous piles. Worn or missing labels make it impossible to determine the exact content of some containers" (Simmons 2000). During the last thirty years, pesticides have seeped into soil, groundwater, and irrigation projects, entering the food chain near many storage areas.

"One nightmare scenario would be some cataclysmic meteorological event that would wash [or] disperse large quantities of DDT or another persistent pesticide into the environment, where the effects could last for many, many years," said Richard Liroff, Washington-based director of the World Wildlife Fund's Alternatives to DDT Project (Simmons 2000).

## PCB RUSTLING IN ZAMBIA

The Environment News Service reported on September 1, 2000, that the Zambia Electricity Supply Corporation (ZESCO), the sole supplier of hydroelectric power in Zambia, had warned the public against unscrupulous people stealing oil containing PCBs from the company's transformers. According to Mellon Chinjila, an official at ZESCO Environment and Social Affairs Unit in Lusaka, the contaminated oil was being sold on the open market in Zambia as cooking oil and skin-lightening lotion. At Nkana Copper Mines, traces of oil thought to contain PCBs were noted around the walls of a building where the theft of thousands of liters of PCB-contaminated oil was reported. Chinjila warned that although this oil may look like genuine cooking oil, it can cause human organ damage and infertility in both men and women. "ZESCO has in the past months been distributing awareness materials on the potential dangers of PCBs to human health and the environment," Chinjila said (Hanyona 2000).

## ORGANOCHLORINES IN SOUTH ASIA

Life-threatening poisons such as DDT, aldrin, chlordane, dieldrin, and heptachlor—all of which are either severely restricted or banned in most countries—continue to be manufactured, stored, used, and traded freely in South Asia, according to an investigative report released by Greenpeace, titled *Toxic Legacies, Poisoned Futures: Persistent Organic Pollutants in Asia* (Hernandez and Jayaraman 1998).

"Asia faces a frightening scenario of historic, current and potential poisoning by the most dangerous variety of persistent poisons. This situation is a result of existing stockpiles of obsolete pesticides, the continuing production of organochlorines and other chemical pesticides and the unmitigated expansion of dirty chlorine-based industries in the region," said Nityanand Jayaraman, a Greenpeace campaigner ("Persistent" 1998).

Greenpeace investigations conducted between April and August 1998 in seven Asian countries, including Bangladesh, India, Nepal, and Pakistan, revealed that stocks of at least five thousand metric tons of obsolete pesticides had been stored under extremely hazardous conditions in more than a thousand sites in Pakistan and Nepal. A sizeable portion of these pesticides arrived as part of aid packages from western countries, and almost all the pesticides were exported by developed nations and India to Pakistan and Nepal.

Chemical corporations whose products were identified in stockpiles in Pakistan and Nepal by Greenpeace investigators included Bayer and Hoechst (Germany); DuPont, Dow Chemical, Diamond Shamrock, and Velsicol (USA); Shell (Netherlands); Sumitomo Chemical and Takeda Chemical (Japan); Rhône-Poulenc (France); Sandoz (Switzerland); ICI (United Kingdom); and Bharat Pulverising Mills (India).

India is among the three remaining manufacturers of DDT in the world, the other two being Mexico and China. India during the late

1990s exported nearly 800,000 kilograms of organochlorine pesticides including aldrin, DDT, BHC, and chlordane to a long list of countries, including countries where their usage is banned. Exports of pesticides that could be branded POPs in the near future such as endosulfan, sodium pentachlorophenate, 2,4-D, and lindane total more than 2 million tons. Some of the pesticides such as aldrin are not permitted even to be manufactured in India. In Pakistan, India, Nepal, and Bangladesh, locally banned or severely restricted pesticides are freely available. Greenpeace found DDT, BHC, dieldrin, and heptachlor openly sold in vegetable markets in Karachi. Hardware stores in New Delhi stock the deadly pesticide aldrin, whose registration was withdrawn there more than two years ago.

"It is unfortunate that while governments in the region are still grappling for ways to dispose of their stockpiles of obsolete imported pesticides, the continuing production and trade of these chemicals goes on unabated. This could only lead to an endless cycle of poisoning whose unwitting and eventual victims are communities and future generations," said Jack Weinberg, international toxics campaigner with Greenpeace. "Governments should aim for an eventual phase-out of such polluting practices and push for international cooperation in developing viable and sustainable non-chemical alternatives" ("Persistent" 1998).

## CONCLUSION: COMING TOGETHER

Friends and neighbors in minority communities across the United States often have found themselves making common cause to evict organochlorine chemicals from their communities. In so doing, they often have found themselves coming together, becoming organized, sharing strategy and tactics, and speaking out against several forms of racism. In this process, leaders have arisen.

One such leader is Mildred Bahati McClain, executive director of the People of Color and Disenfranchised, who describes herself as a "mother, grandmother and a steward of my community," (McClain, n.d.). McClain, who lives in Savannah, Georgia, also is a singer who has enthralled many audiences.

McClain said of her childhood:

I grew up in Savannah smelling the stench from the Union Camp Paper Company thinking the smell was a natural part of life. My mother's friends who lived near Union Camp were always complaining of headaches, bad skin rashes, kidney problems and severe cases of asthma. . . . I am frustrated because we are clearly being discriminated against. Many think we are crying wolf, but I ask you to come to our neighborhoods and see for yourself. You will be in shock. (McClain, n.d.)

McClain has become a major organizer of resistance to organochlorine-manufacturing plants in minority communities, notably in the southern regions of the United States. Along the way, she and her supporters have made some political points about racism as well. McClain tells the following story about a meeting in New Orleans:

On Sunday morning . . . several Cherokee Indians were treated miserably in a Shoneys restaurant. They were ignored and then received only partial and reluctant service. When this was made known to those attending the conference, participants quickly organized a march involving over 200 people. We went to the restaurant to demand an apology. In a very moving meeting, the tearful manageress apologized and said it would never happen again. It felt really good on that march. We had Native Americans, African Americans, Asian Americans, Hispanic Americans and White Americans marching together. It felt like a microcosm of what we had to do. Just after we got back to the hotel, the heavens opened. Native Americans saw this as a "cleansing." (Work on Waste 1996)

## REFERENCES

Allen, Joe. 1995. "Malathion in Mission." *Circle* (Minneapolis), April.

Bowermaster, J. 1993. "A Town Called Morrisonville." *Audubon,* July/August, 42–51.

Brown, David. 2000. "Defoliant Connected to Diabetes." March 29. http://irptc.unep.ch/pops/newlayout/press_items.htm.

"Celebrities to Tour 'Cancer Alley,' Louisiana; Alice Walker, Alfre Woodard, and Mike Farrell among Speakers at National Town Meeting on Environmental Justice." No date. http://www.greenpeaceusa.org/toxics/canceralleytour/celebritytour.htm.

Costner, Pat. 1995. *PVC: A Primary Contributor to the U.S. Dioxin Burden; Comments Submitted to the U.S. EPA Dioxin Reassessment.* Washington, D.C.: Greenpeace U.S.A.

Costner, Pat, and Joe Thornton. 1989. *We All Live Downstream: The Mississippi River and the National Toxics Crisis.* Washington, D.C.: Greenpeace.

Cray, Charlie. 1996. "Hundreds Oppose Shintech Proposal in Louisiana: Citizens and Other Interest Groups Cite Health Concerns." Greenpeace. December 9. http://lists.essential.org/1996/dioxin-l/msg00752.html.

Cray, Charlie, and Monique Harden. 1998. "PVC and Dioxin: Enough Is Enough." *Rachel's Environment and Health Weekly* 616, September 18. http://csf.colorado.edu/envtecsoc/98/0285.html.

"Dioxin Deception: How the Vinyl Industry Concealed Evidence of Its Dioxin Pollution." 2001. Greenpeace. March 27. http://www.greenpeaceusa. org/toxics/dioxin_deceptiontext.htm.

Duchin, Lelanie. 1997. "Greenpeace's Secret Sampling at U.S. Vinyl Plants: Dioxin Factories Exposed." Greenpeace. April. http://www.green peace.org/~toxics/reports/reports.html.

Hanyona, Singy. 2000. "Zambia Struggles to Control Toxic PCBs." Environment News Service, September 1, 2000. http://www.repp.org/ discussion/stoves/200009/msg00001.html.

Hernandez, V., and N. Jayaraman. 1998. *Toxic Legacies, Poisoned Futures: Persistent Organic Pollutants in Asia.* Washington, D.C.: Greenpeace.

Johansen, Bruce E. 1993. *Life and Death on Mohawk Country.* Golden, Colo.: North American Press/Fulcrum.

Krook, L., and G. A. Maylin. 1979. "Industrial Fluoride Pollution: Chronic Fluoride Poisoning in Cornwall Island Cattle." *Cornell Veterinarian* 69, suppl. 8: 1–70. http://www.ncbi.nlm.nih.gov/htbin-post/Entrez/ query?uid = 467082&form = 6&db = m&Dopt = r.

Kurtz, Howard. 2001. "Moyers's Exclusive Report: Chemical Industry Left Out." *Washington Post,* March 22.

"Louisiana's Cancer Alley: An International Threat." 1999. Environment News Service, March 5. http://ens.lycos.com/ens/mar99/1999L-03- 05-09.html.

"Louisiana Town Residents Exhibit Dangerous Levels of Dioxin." n.d. Greenpeace USA. http://www.greenpeaceusa.org/features/mossville text.htm.

MacEachern, Frank, and Rachele Labrecque. 2001. "Clean-up Causes Health Fears." *Cornwall* (Ontario) *Standard-Freeholder,* July 13. http://www.standard-freeholder.southam.ca.

McClain, Mildred. n.d. Food First Economics Bus Tour: More Testimonies. http://www.foodfirst.org/bustour/testimonies2.html.

Moyers, Bill. 2001. "Trade Secrets: A Moyers Report. Program Transcript." Public Broadcasting Service, March 26. http://www.pbs.org/trade secrets/transcript.html.

*National Law Journal.* 1992. Special Issue: Unequal Protection: The Racial Divide in Environmental Law, September 21.

Negoita, S., L. Swamp, B. Kelley, and D. O. Carpenter. 2001. "Chronic Diseases Surveillance of St. Regis Mohawk Health Service Patients." *Journal of Public Health Management Practice* 7, no. 1: 84–91.

"Persistent Organic Pollutants in Asia: An Ongoing Disaster." 1998. Greenpeace, November 10 http://www.greenpeace.org/pressreleases/toxics/ 1998nov10.html.

Pierson, Nova. 2000. "Toxic Travel Fears." *Calgary Sun*, January 17.

"Public Notice, Air Permit Application, Shintech Corporation, December 1997, and Shintech Application [to the EPA] to Discharge Process Wastewater." No date.

Rusnell, Charles. 2000. "U.S. Military Wastes Entering Canada; Ottawa Concerned with Political Fallout, Document Shows." *Edmonton Journal*, March 31.

Schettler, Ted. 2000. "Statement of Ted Schettler, MD, Physicians for Social Responsibility." Press Conference, Willard Hotel, Washington, D.C., September 7. http://www.psr.org/trited.html.

Schettler, Ted, Gina Solomon, Maria Valenti, and Anne Huddle. 1999. *Generations at Risk: Reproductive Health and the Environment*. Cambridge, Mass.: MIT Press.

Seely, Hart. 2001. "Toxins Remain 18 Years Later: Landfill near Massena Polluting Water Where Mohawk Children Played." *Syracuse Post-Standard*, June 24.

Sengupta, Smini. 2001. "A Sick Tribe and a Dump As a Neighbor." *New York Times*, April 7. http://www.nytimes.com/2001/04/07/nyregion/07-MOHA.html.

"Shintech: The Battle Continues." 1999. *E: The Environmental Magazine*, March–April. http://www.emagazine.com/march-april_1999/0399up dates.html.

Sierra Club. n.d. "Stories from the Field: Corporate Pollution: Shintech and Louisiana; Eyes of the World on 'Cancer Alley.'" http://www.sierra club.org/toxics/resources/shintech.asp.

Simmons, Ann M. 2000. "Tanzania Begins to Deal with Toxic Wastelands, Pesticides; With No Sound Disposal Method, Stockpiles Keep Growing. Experts Warn That Unless Quick Action Is Taken, the Situation Could Be Catastrophic." *Los Angeles Times*, March 30. http://irptc.unep.ch/pops/newlayout/press_items.htm.

"State, G.M. Talking So Lawsuit Is Set Aside." 2001. Associated Press, June 11. http://syracuse.com/newsflash/index.ssf?/cgi-free/getstory_ssf.cgi?n0505_BC_NY—contamination&&news&newsflash-newyork-syr.

"State, Mohawks Threaten Lawsuit Unless G.M. Cleans Up." 2001. Associated Press, March 22. http://www.topica.com/lists/SSVOP.

Thomas, Katie. 2001. "Toxic Threats to Tribal Lands." *Newsday*, March 25. http://www.newsday.com/coverage/current/news/sunday/nd8399.htm.

"Toxic Waste in Japan: The Burning Issue." 1998. *Economist* (London), July 25, 60.

"U.N. Agency Calls for Faster Disposal of Toxic Pesticide Waste Stocks." 2001. Press release, U.N. Food and Agriculture Organization. May 9. www.fao.org.

U.S. Environmental Protection Agency. 1998. *Federal Register* 63, no. 83

(April 30): 23785–86. http://www.epa.gov/fedrgstr/EPA-AIR/1998/April/Day-30/a11512.htm.

"U.S. Government Turns Its Back on Dioxin Elimination in Global Pollution Treaty." 1999. Greenpeace. Accessed at http://www.greenlink.org/public/hotissues/dioxin.html.

Versar, Inc. 1983. *Exposure Assessment for Incidentally-Produced Polychlorinated Biphenyls (PCBs)*. Draft Final Report, vol. 3, Appendix B: "Organic Chemicals Possibly Associated with Incidentally Produced PCBs." Report by Stanley Cristol, University of Colorado. Appendix C: Chemical by Chemical Exposure Summary Reports. U.S. EPA Contract No. 68-01-6271. August 15.

"With Public's 'Right-to-Know' in Jeopardy, Greenpeace Kicks Off Bus Tour of Louisiana's Worst Chemical 'Hot Spots.'" 1999. Greenpeace, June 22. Accessed at http://www.commondreams.org/pressreleases/june99/062299e.htm.

Work on Waste. 1996. "The Third Citizens' Conference on Dioxin and Other Synthetic Hormone Disrupters, March 15–17, Baton Rouge, Louisiana." http://www.workonwaste.org/wastenots/wn357.htm.

# 5

# Belugas with Tumors: The Toxic
# Toll on Animals

We live in a world that now offers no refuge from the synthetic chemical effluent of industry—a fact that is attested, again and again, by levels of various POPs in the body fat of mammals, such as polar bears, which live thousands of miles from the factories producing the chemicals and most of the people who use them. Meat-eating birds, such as bald eagles and vultures, also are unwilling worldwide witnesses to the spread of DDT, PCBs, dioxins, and other synthetics into the entirety of the biosphere.

Such is the ecological state of a world in which vultures (in India) die from eating DDT-contaminated cattle carrion. In today's world, PCB toxicity levels in whale meat are causing some of its major consumers, the Japanese, to become leery of eating it. Toxicity, more than any human "green" consciousness, thus may be "saving" the whales, at least for the time being. Whales, like all mammals, pass their toxic burden to offspring, in concentrated form, imperiling each successive generation with greater potential for cancers, reproductive failure, and other maladies that follow from increased POP toxicity. We live in an ecological world in which beluga whales grow cancerous tumors, a world in which, sometimes, not even an informed observer can tell which polar bears are male and which are female.

An extensive body of scientific evidence has documented the devastating toll of persistent organic contaminants on wildlife. In many parts of the world, wild species show signs of disrupted sexual development and a diminished ability to reproduce. Some sensitive species have disappeared altogether because of total reproductive failure linked to some of the dirty dozen chemicals on the POPs list.

One well-informed observer believes that "Evidence is conclusive that dioxin (TCDD) causes cancer in animals" (Allsopp, Costner, and Johnston 1995).

In laboratory animals, exposure to dioxins, particularly 2,3,7,8-TCDD (a form of dioxin), has been associated with a large number of toxic effects. Some of these effects have occurred at very low doses. For instance, exposure of monkeys to only 5 parts per trillion of 2,3,7,8-TCDD caused impaired neurological development and endometriosis (Rier et al. 1993). Pregnant rats receiving a single small dose of 2,3,7,8-TCDD on day fifteen of pregnancy had male offspring that appeared normal at birth but at puberty were "de-masculinized, with altered reproductive anatomy, reduced sperm count, feminized hormonal responses, and feminized sexual behavior" (Mably et al. 1991; Thornton 1997). In 1992, researchers at the University of Wisconsin reported that low-level prenatal exposures to dioxin feminized the behavior of male rats during adulthood and sharply reduced their production of sperm (Mably et al. 1992a).

Very low doses of dioxin also have produced immune-system changes in rats and monkeys (Hong, Taylor, and Abanour 1989; Neubert et al. 1992; Yang, Lebrec, and Burleson 1994). Reproductive function may be disrupted by exposure to PCBs in animals, especially mammals (including humans). Female rhesus monkeys exposed to PCBs "have alterations in menstrual cycles (e.g., duration and bleeding), decreases in fertility, increased abortions and reductions in the number of conceptions" (Arnold, Mes, and Bryce 1990; Barsotti, Marlar, and Allen 1976).

PCBs have been implicated in the disappearance or decline of several animal species in the United States and Europe. Mink, for example, began disappearing from the shoreline of the Great Lakes during the mid-1950s. Despite restrictions on DDT, PCBs, and other POPs, mink have not yet returned. At the same time, British researchers have linked PCBs to the decline of otters in Britain and Europe during the 1950s (Lopez-Martin, Ruiz-Olmo, and Minano 1994). Mink are very sensitive to reproductive impairment induced by prenatal organochlorine chemical exposure.

Dioxins and PCBs are toxic to the immune systems of many animals. Many studies support evidence that exposure to extremely low doses of dioxin increases susceptibility to bacterial, viral, parasitic, and neoplastic diseases (U.S. EPA, 1994). Many of these chemicals also act as endocrine disrupters, some of whose effects are alligators born with abnormally small penises and birds with crossed beaks. Endocrine disrupters confuse the body's sexual identity. Effects of endocrine disruption include interrupted sexual development, thyroid-system disorders, inability to breed, reduced immune-system response, and abnormal mating and parenting behavior. Some species for which abnormalities have been reported include terns, gulls, harbor seals, bald eagles, beluga whales, lake trout, panthers, alligators, turtles, and others.

Bald eagles around the Great Lakes and in the Columbia River basin in Washington State were unable to reproduce successfully after they were fed local fish for two or more years. The bodies of the fish-eating bald eagles have been contaminated with enough DDT, PCBs, and chlordane to compromise their reproductive capacities, despite the fact that all these chemicals had been banned for twenty years by the time measurements were taken during the early 1990s (Montague 1993). The chemicals continued to be imported into the United States from countries still manufacturing them, and existing stocks are very persistent. "PCBs will be around over geologic time"—thousands of years—one report asserts (Colborn et al. 1993, 378).

In all mammals, male and female reproductive organs are particularly at risk during fetal development after maternal exposure to organochlorines. The reproductive organs at risk are the mammary glands, uterus, cervix, and vagina in women and the seminal vesicles, prostate, epididymides, and testes in men. The skeleton, thyroid, liver, kidney, and immune systems are affected in both sexes.

A fetus grows rapidly and is therefore very susceptible to damage from toxic insults. Damage to the dividing cells of a developing fetus often causes severe deformity. Not only is the fetus growing rapidly, but it also has little body fat, no protective reservoir for fat-soluble toxins such as organochlorines, as in adults. Enzyme systems are immature, so a fetus can detoxify toxic substances at a rate too low to compensate for exposure. Permanent effects are most likely to arise from exposure during this point in the life cycle. These effects may be immediately apparent, such as deformities at birth. Many effects also may not become obvious until the offspring exposed prenatally reaches maturity (Colborn et al. 1993).

## PITY THE POLAR BEAR

Pity the polar bear, whose food supply (mainly seals) now arrive less often on receding ice, and whose body fat has become contaminated, more so generation by generation, with chemical toxins such as dioxins and PCBs that bioaccumulate along its food chain. The polar bears are subject to the same environmental stresses as the Inuit with less ingenuity to outwit the pending environmental apocalypse in the Arctic. We may be able to count the polar bear generations to extinction on one set of human fingers if production and use of persistent organic pollutants is not stopped soon.

As with Inuit mothers, polar bears' offspring are one step of biomagnification along the food chain:

Stiff and lean after six months curled in a den, a female polar bear squeezes herself out of her winter home. Two small cubs emerge tentatively at her heel for their first view of the world beyond a snow cave. Entirely dependent on their mother, the cubs follow obediently. Having used up most of her fat stores, the female scans the sea ice below and ponders a meal of seal blubber. But her cubs are not yet ready to travel, and her milk will have to sustain them for some time to come. The milk is rich and nourishing but today it also harbors a threat. The seals that the mother has feasted on in the past, and will need to eat again soon, are tainted by chemicals from lands far beyond her sea-ice domain. The chemicals that bind to the fat of the seals have accumulated in her own fat stores. Unwittingly, the mother passes the toxins to her young in her fat-rich milk, with effects that are still unclear. ("Persistent Organic" n.d.)

Polar bears sit at the top of a major food chain in the Arctic, an area that has become a major sink, or repository, for POP contamination generated at lower latitudes. If an animal's milk-fat content is higher, the magnification of POPs such as dioxin and PCBs is higher; POP concentrations in dolphins accelerate more quickly through the generations than for human beings, because their milk is richer in fat-harboring contaminants. Polar bears and seals, as well as other Arctic or Antarctic mammals, feed their offspring a fat-contaminated toxic cocktail of POPs.

Increasing POP levels among some polar bears already have, in some cases, transformed their sexual organs, confusing male and female. Scientists on the Svalbard islands have found that more than one in a hundred of the islands' polar bears are hermaphroditic. The condition, in which an animal possesses the reproductive organs of both sexes, afflicts wildlife in various parts of the world.

It may be caused by exposure to chemicals that affect the endocrine system.

PCBs probably are responsible for the Svalbard bears' physiological sexual confusion. The same pollutants also are compromising the bears' immune systems. This phenomenon was unknown a decade ago on the islands, which lie between Norway and the North Pole. By the year 2000, 1.2 percent of Svalbard's bears, which total about 3,000, had two sets of sexual organs (Kirby 2000). Per Kyrre Reymert, of the Svalbard Science Forum, told BBC News Online: "Tests have been conducted on 40 bears, and these are now being analyzed. There is a very short and simple food chain here—plankton, fish, seals, and finally the bears themselves. So it is fairly easy to track PCBs and other pollutants" (Kirby 2000).

Elizabeth Salter, of the World Wildlife Fund, told British Broadcasting Corporation News Online: "It seems to be the female bears on Svalbard that are acquiring male genitalia, a penis-like stump. . . . what's happening to the bears is happening to gulls in the Arctic, too, because of PCBs, DDT and dioxins" (Kirby 2000). Per Kyrre Reymert fears there will be more encounters between humans and bears. "Climate change is likely to reduce the amount of ice around Svalbard. With less ice, there will be more bears coming ashore. And they will be hungry," Reymert said (Kirby 2000).

Polar bears are among several predatory marine mammals that consume high levels of organochlorines. Organochlorine chemicals including DDT, PCBs, chlordanes, HCHOs, and PCCOs (polychlorinated camphenes, e.g., toxaphene) have been widespread in the Arctic, affecting fur seals (*Callorhinus ursinus*), ringed seals (*Phoca hispida*), hooded seals (*Cystophora cristata*), bearded seals (*Erignathus barbatus*), walrus (*Odobenus rosmarus divergens*), beluga (*Delphinapterus leucas*), porpoises (*Phocoena phocoena*), narwhal (*Monodon monoceros*), and polar bears (*Ursus maritimus*). All are relatively high on their respective food chains and very susceptible to bioaccumulation of these organochlorines. In the Wadden Sea, reproductive failure of the common seal (*Phoca vitulina*) and subsequent population decline has been attributed to PCBs (Norstrom and Muir 1994).

## HEAVY METALS IN ALASKAN REINDEER

The Seward Peninsula of Alaska has been extensively mined for the last hundred years. The mining of cadmium- and lead-bearing ores and the widespread use of these two elements for industrial purposes have significantly increased environmental contamination over the last century. These elements are readily absorbed by plants that are in turn eaten by ungulates, concentrating in liver, kidney, and muscle tissue. In some cases, health officials have recommended against consumption of reindeer meat, after learning that contamination from weapons testing, accidental pollution, and illegal dumping may have found their way to the lichens of Northwestern Alaska. The same contamination also has accumulated in reindeer and caribou tissue ("Heavy Metal" 2000).

Many people on Alaska's Seward Peninsula live a subsistence lifestyle with most of their food coming from local plants and animals. The incidence of cancer and other diseases appears to be rising among the native people in this region. "Though the link between epidemiology and environmental factors potentially involves a multitude of causal relationships, the people in the villages are particularly concerned that contaminants from air pollution, mining operations, and dumpsites are concentrating in the tissue of subsistence animals and pose a health risk" ("Heavy Metal" 2000). The Reindeer Research Program detected high levels of cadmium and lead in several species. At similar concentrations, consumption of forty to sixty grams of meat per week would exceed the recommended intake rate ("Heavy Metal" 2000). Similarly, reindeer and pike in northern Sweden were found to be contaminated with radioactivity after the nuclear accident at Chernobyl in the Ukraine (Forberg, Tjelvar, and Olsson 1991; Skogland 1987).

## BELUGAS WITH TUMORS

While levels of various industrial chemicals (notably PCBs) in the St. Lawrence River have declined since the 1960s, Beluga whales still show high levels of toxicity, with higher concentrations in younger whales. These findings indicate that these contaminants are being transferred across the placenta from mother to offspring. The belugas' fatty breast milk has been bioaccumulating toxins across generations long after human injection of POPs into the river decreased substantially. In this way, young belugas acquire a toxic load much higher than their parents' in the absence of new PCB contamination. Researchers found one beluga in the St. Lawrence that had 500 p.p.m. of PCBs in its body fat, ten times the concentration required to classify it as toxic waste under Canadian environmental law (Colborn, Dumanoski, and Myers 1996).

In the St. Lawrence River, beluga whales suffer from an astonishing list of afflictions—several kinds of cancer, twisted spines and skeletal disorders, ulcers, pneumonia, bacterial and viral infections, thyroid abnormalities—seldom if ever seen in belugas living in less-polluted water. Ongoing research on this population indicates that widespread hormone disruption is undermining reproduction and preventing recovery of the population (World Wildlife Fund 2000).

During the past decade, scientists also have found that contaminants such as DDT, PCBs, and dioxins weaken the immune systems of marine mammals. This evidence now makes it appear likely that contaminant-induced immune suppression may have contributed to the dramatic marine epidemics that killed thousands of seals, dolphins, and porpoises in the late 1980s and early 1990s. The dramatic die-offs hit populations in the Baltic and North Seas, the Mediterranean, the Gulf of Mexico, the North Atlantic, and the eastern coast of Australia. The carnage extended even to seals in Siberia's Lake Baikal.

## MALE FISH SECRETING FEMALE HORMONES

During the 1980s, John Sumpter, a professor of biochemistry at Britain's Brunel University, found male fish secreting female hormones in the River Lea, which runs through an industrialized section of North London. The bodies of the male fish were creating large amounts of the female egg-yolk protein vitellogenin. Usually males do not make any of this substance. "The levels of vitellogenin in the blood of the males was extraordinarily high, increased by about 100,000 times. The levels reached are those we would normally expect to find in a fully mature female trout which was making lots of eggs," said Sumpter (Cadbury 1997, 114). Sumpter and his colleagues analyzed sewage and other discharges into the River Lea and were among the first to link synthetic estrogens to the feminization of male fish. Female mollusks (e.g., snails and mussels) have become male after exposure to endocrine-disrupting chemicals (a condition known as imposex).

When caged trout were placed downstream from sewage treatment plants in several British rivers, the males developed excessive levels of vitellogenin in their blood (Purdom et al. 1994; Sumpter 1995). Every sewage-treatment plant in England caused this estrogenic effect within two to three weeks' exposure. The British researchers found that several common industrial chemicals produce vitellogenin in male fish under laboratory conditions: octylphenol and nonylphenol (both alkyl phenols, used in detergents, toiletries, lubricants, and spermicides); bisphenol-A (used in polycarbonate plastics); DDT (the common pesticide); and Arachlor 1221, one of the 209 varieties of PCBs (Sumpter and Jobling 1995). The same researchers also tested chemical mixtures and found that the mixtures were more powerful progenitors of vitellogenin than any of the individual chemicals alone. These studies also documented bioaccumulation of the chemicals. Thus, a weakly estrogenic chemical could eventually build to a level capable of producing vitellogenin.

Sumpter and Jobling also tested for estrogenic effects in fish across species. They concluded that "Most evidence supports the idea that if a chemical is estrogenic in one species, it will be in all others" (Montague 1997a). Late in 1996, researchers in the United States published studies confirming that sewage-treatment effluent there also can cause the same effects in fish living downstream (Folmar et al. 1996).

Scientists examined male carp from five locations in the Mississippi River downstream from a Minneapolis sewage-treatment plant, as well as from a tributary, the Minnesota River, which bears intense agricultural chemical runoff. They found that carp living near the Minneapolis sewage-treatment plant showed "a pronounced estrogenic effect," producing vitellogenin along with reduced levels of testosterone (Folmar et al. 1996, 1096). In addition, the site with the highest level of dissolved pesticides in the United States (as tested by the U.S. Geological Survey), the Platte River at Louisville, Nebraska, had the lowest estrogen-to-testosterone ratio in the United States (Montague 1997a).

Infusion of dioxins and furans into the marine environment from neighboring industrial plants caused a particularly high level of contamination among crabs in a Norwegian fjord. Similar levels of contamination have been noted in Newark Bay, New York Bight, and Tokyo Bay. For freshwater fish, studies on various rivers showed that in Spain, heptachlor epoxide exceeded World Health Organization (WHO) standards; in Australia, PCBs and chlordane also exceeded WHO standards. Fish taken from the St. Lawrence River and rivers in British Colombia had higher than usual concentrations of dioxins and furans for their areas and the world as a whole. In southern Taiwan, fish taken from a river and culture ponds exceeded dioxin standards as a result of industrial activities in the area.

## POPS AND THE DECLINE OF FROG POPULATIONS

Pesticide residues drifting up Sierra Nevada slopes from California's agricultural valleys have been linked to drastically declining frog populations there. Researchers from the U.S. Geological Survey and the U.S. Department of Agriculture assert that the pesticides disrupt an enzyme that regulates the frogs' nervous systems. Frog populations have been reported to be in decline (with an upsurge in deformities, such as extra or missing limbs) in many areas, but this study is among the first to assay a cause for a specific amphibian decline. Commented biologist Jessa Netting, writing in *Nature,* "Amphibians, with their moist, sensitive skins, unprotected eggs, and semi-aquatic lifestyle, have long been viewed as biological indicators of environmental health" (Netting 2000, 760). Ultraviolet radiation and parasites also are suspected factors in other frog-population declines.

Industrial toxins, notably agricultural herbicides, are one of several factors contributing to widespread frog population declines and deformities worldwide. *Froglog,* published by the Species Survival Commission of the Declining Amphibian Populations Task Force of the World Conservation Union has reported that the 1996 Red List of Threatened Animals (published by the International Union for Conservation of Nature) listed 156 amphibian species as extinct, critical, endangered, or vulnerable to extinction. This represents 25 percent of all the amphibians on Earth (Halliday 1997). The Nature Conservancy, a U.S. organization, in 1996 surveyed the status of 20,481 species of plants and animals in the United States and reported that 37.9 percent of U.S. amphibians are in danger of becoming extinct (Dicke 1996).

Researchers at Widener University in Chester, Pennsylvania, and at Benedictine College in Atchison, Kansas, have shown that acid rain can stress frog populations by harming their immune systems

(Brodkin and Simon 1997). According to one observer, frogs raised in water with a pH of 5.5 had significantly more bacteria in their spleens and a significantly higher death rate than frogs raised in waters with a pH of 7.0. The researchers attribute the increased numbers of bacteria to reduced efficiency of bacteria removal by white blood cells—part of the frogs' immune defenses (Chow 1997; Montague 1998).

High levels of organochlorine pesticides have been correlated with reduced frog populations in several parks and wildlife reserves along the northern edge of Lake Erie. At Point Pelee National Park in Canada, only five frog species remain; DDT residues in these frogs average 5,000 to 47,000 micrograms of DDT per kilogram of body weight (Russell and Hecnar 1996; Russell et al. 1995, 1997).

During the early 1990s, researchers discovered many frogs and toads with missing back legs in ponds and ditches exposed to pesticide runoff in the St. Lawrence River valley in Quebec, Canada. Twelve percent of the frogs examined in one study had hind-limb deformations that are otherwise "virtually unknown" among amphibians in the wild (Ouellet et al. 1997, 95). In 1997 the Australian government banned eighty-four herbicide products for use near water because of their harmful effects on tadpoles and frogs (Tyler 1997). Researchers in Sri Lanka have reported that frogs are nearly absent on tea plantations where herbicides are used intensively. Sri Lanka frog populations rebounded quickly after spraying stopped. "Conversion to pesticide-free tea production in this region has contributed greatly to the re-establishment of populations of local frogs," according to this study (Senanayake et al. 1997, 2).

Peter Montague has observed that "All of the banned products contain Monsanto's glyphosate as the active ingredient. However, the harmful component appears to be not the glyphosate itself but an 'inert' ingredient, a detergent or wetting agent added to the herbicides so that droplets of liquid spread out and cover the target leaves. Detergents interfere with the ability of frogs to breathe through their skins, and tadpoles to breathe through their gills" (Montague 1998). Michael J. Tyler of the Department of Zoology at the University of Adelaide, Australia, says, "Although the herbicide [glyphosate] is claimed to be 'environmentally friendly,' it is clear that users have been lulled into a false sense of security" (Montague 1998).

# ATRAZINE, FROGS, AND SEXUAL DYSFUNCTION

Atrazine (2-chloro-4-ethytlamino-6-isopropylamine-1,3,5-triazine), the most frequently used herbicide in the United States, has been associated with a wide array of sexual abnormalities in frogs by a study published in the *Proceedings* of the U.S. National Academy of Sciences. Tyrone B. Hayes, a specialist in the hormone systems of amphibians at the University of California at Berkeley, led the team who conducted the study. Laboratory experiments there suggest that even at very low levels, exposure to atrazine can disrupt hormones and alter the sexual development of male African clawed frogs.

The study was published as the U.S. Environmental Protection Agency entered the final phase of its special review of the chemical, begun during 1994 after atrazine was found to be widespread in U.S. drinking water. Although atrazine has not been proven as dangerous to humans as to frogs, studies compiled by the U.S. Geological Survey have detected levels as much as 2,300 p.p.b. in drinking water consumed near some agricultural areas. The EPA has attempted to limit drinking-water concentrations to 3 p.p.b.

The Hayes team's findings indicate that atrazine is a powerful endocrine disrupter in frogs. Exposure of tadpoles to water with 0.1 p.p.b. of atrazine, or 30 times less than the amount legally permitted in drinking water, induced male frogs to develop both testes and ovaries. "The atrazine turns on an enzyme called aromatase," Hayes said. "This converts testosterone to estrogen" (Green 2002, A-16). Roughly 20 percent of the frogs that the Hayes team tested showed hermaphroditic characteristics, a number the team has discovered translates into 40 percent of the population's males.

"We have males with up to six testes, for example," Hayes said. "We have animals with three testes and three ovaries" (Nelson 2002, A-20). When exposed to 1 part atrazine per billion parts water, Hayes said 90 percent of the male frogs suffered a deficient voice box. Because frogs sing to attract mates, the deficiency poses a major handicap. "No song, no offspring," Hayes said (Nelson 2002, A-20). At 25 p.p.b. of atrazine, a level frequently observed in rainfall and surface waters in some midwestern agricultural areas, Hayes and colleagues found that sexually mature males suffered a tenfold decrease in testosterone, a decline that Hayes said amounts to "functionally castrating the animals" (Nelson 2002, A-20).

Sensitivity of frogs to atrazine at levels of a few p.p.b. is important because intense agricultural use of the chemical has caused it to become part of the natural background. Atrazine was, by the year 2000, being used in eighty countries around the world. Atrazine has generally been regarded as relatively safe because of its short half-life and low rate of biomagnification. In human adults, a sketchy test record has indicated that it causes abnormalities only at very high doses (Hayes et al. 2002, 5476).

Hayes and his colleagues found atrazine present at a saturation of "several parts per million in agricultural runoff, and [as much as] 40 parts per billion in precipitation" (Hayes et al. 2002, 5476). By contrast, according to tests by Hayes and colleagues, "by immersion throughout larval development," deformations of male frogs' sexual organs (hermaphroditism and demasculinization) began at any level more than 1 p.p.b., a level that is now present in precipitation even in areas where it is not used (Hayes et al. 2002, 5476). According to Hayes et al., at a level of 25 p.p.b., male *Xenopus laevis* (African clawed frogs) suffered a 10-fold decrease in testosterone levels.

Syngenta, the company that manufactures most of the atrazine sold in the United States, funded some of Hayes's early research under pressure from the EPA but rejected its results. Syngenta toxicologist Tim Panoor said other studies commissioned by the firm produced contrary results. "No conclusions can be drawn," he said (Green 2002, A-16). Atrazine was first synthesized in Switzerland in 1955 and was licensed in the United States during 1959. Of the 75 million pounds now used in the United States every year, most are applied in the Midwest on corn, sorghum, and sugarcane crops (Green 2002, A-16). Atrazine has been banned in Switzerland, Sweden, Germany, Italy, Norway, France, and Belgium (Dalton 2001, 665).

"This is not a worst-case scenario where animals are exposed to mega-doses," commented Louis Guillette, a zoologist at the University of Florida in Gainesville, who has studied the effects of persistent organic pollutants in alligators. "These are concentrations we know wild amphibians are exposed to" (Dalton 2002, 665).

Scientists debating whether high rates of deformities in frogs are the result of parasites or pollution may both be right, according to a study published July 9, 2002, in the *Proceedings of the National Academy of Sciences*. Penn State University biologist Joseph Kiesecker wrote that infection with trematode worms, a common parasite, actually causes the deformities in wood frogs but that the deformity rates were substantially higher in areas where infected frogs were exposed to pesticide runoff" ("Links Found" 2002). Kiesecker said the chemicals may lower frogs' immunity to infection, or pesticide runoff may boost the population snails that transmit the trematodes to frogs.

Kiesecker's study examined frogs in six central Pennsylvania ponds. Some had pesticide runoff, and some did not. Results from the wild were compared to frogs in the lab. In ponds without pesticide runoff, deformities occurred in 5 percent to 10 percent of trematode-infected frogs. In ponds where runoff was present, deformity rates ranged from 20 percent to 30 percent ("Links Found" 2002). In the wild, deformities occurred only where the parasite could get at the frogs—confirming the parasite's essential role in the malformations, Kiesecker reported in the *Proceedings of the National Academy of Sciences.*

"If it's true that commonly used pesticides compromise the immune system of a vertebrate organism, which is what these findings suggest, then we're looking at a much bigger problem than deformed frogs," said David Gardiner of the University of California at Irvine (Souder 2002, A-9). The pesticide concentrations used in this study fell below government-recommended levels for safe drinking water. Among them was atrazine, the most heavily used agricultural herbicide in the United States.

According to a report in the *Washington Post*, "In the lab, Kiesecker found much higher rates of parasitic infection in tadpoles exposed to pesticides, along with a matching reduction in white blood cell production—an indication of a weakened immune response. Frogs exposed to pesticides were also smaller and developed more slowly. Kiesecker said inhibited growth rates might also contribute to deformities seen in the field by creating a longer 'window' when parasitic infection could affect limb development" (Souder 2002, A-9).

"The original purpose of these experiments was to study how disease in aquatic systems impacts growth and development," Kiesecker said. "To be honest, we were quite surprised to see limb deformities and this strong pattern of interaction between parasites and pesticides" (Souder 2002, A-9). "Amphibians have become an important model system," Kiesecker said. "We have to consider that factors that influence infection rates in frogs may also play a role in human diseases" (Souder 2002, A-9).

Kiesecker's results contradicted tests sponsored by Syngenta Crop Protection Inc., a manufacturer of atrazine, which failed to support a conclusion that the chemical is causing widespread deformation of frogs. The studies were unable to replicate the results of earlier research that alleged that atrazine may affect the larynx and sexual development of African clawed frogs. The studies were conducted by a panel of eight university scientists convened by Ecorisk—a global network of environmental scientists—to examine the environmental effects of atrazine ("UV Radiation" 2002).

"As research on this issue continues, one thing is certain," said Ronald Kendall, director of the Institute of Environmental and Human Health at Texas Tech University and chair of the science panel. "No conclusions can be drawn at this time on atrazine and its purported effect on frogs. We must get the science done in order to have the facts to make sound conclusions" ("UV Radiation" 2002).

## DIOXINS AND ENDOMETRIOSIS

Permanent exposure to low levels of dioxin can cause endometriosis in monkeys (Montague 1993). Endometriosis is a painful disease of the tissues lining the uterus, which often results in sterility; it presently afflicts 5 to 9 million women in the United States. Male rats exposed to PCBs prior to birth have disturbed thyroid hormones; as a side effect, these rats have small testicles and reduced sperm counts as adults (Montague 1993). Exposure of rodents to endocrine-disrupting chemicals "can cause them to undergo puberty at an early age and can cause persistent estrus [being 'in heat'] for a . . . prolonged time," according to the Weybridge Report, a key compilation of pesticide effects on health (Montague 1997b).

Male rodents exposed to chemicals that interfere with androgens (male sex hormones) can be born with hypospadias (a birth defect of the penis) and cryptorchidism (undescended testicles) (Montague 1997b, part 1, p. 29). The Weybridge Report attributes these effects to Vinclozolin, a powerful antiandrogenic pesticide. Remarks Peter Montague: "In the United States today, Vinclozolin is legal for use on cucumbers, grapes, lettuce, onions, bell peppers, raspberries, strawberries, tomatoes, and Belgian endive. The U.S. Environmental Protection Agency has no published plans for banning Vinclozolin" (Montague 1997b).

Furthermore, according to Peter Montague, "There is evidence from a number of animal species that sex steroids (for example, estrogen and testosterone) exert long-term effects on the size of the thymus and on the immune system in general" (Montague 1993). In humans, the thymus is an organ, just above the heart, that produces T cells, which are crucial actors in the immune system" (Montague 1997b).

## CONTAMINATION IN NORTHERN FUR SEALS

The breeding rookeries for more than 70 percent of the world's population of northern fur seals (*Callorhinus ursinus*), in the Bering Sea on the two largest Pribilof Islands (St. Paul and St. George, Alaska) offer evidence of reproductive problems caused by organochlorines. The St. Paul breeding population appears to have stabilized after a decline of about 7 percent per year between 1975 and 1983. However, on St. George, pup production declined at an annual rate of about 6 percent from 1973 to the middle 1990s (Beckmen, Krahn, and Stott 1997). The Aleut village culture on these islands is tightly bound to a subsistence harvest of the northern fur seal.

Pregnant northern fur seals accumulate organochlorines during winter feeding before their June migration to the Pribilofs to give birth. While pregnant, females transfer organochlorine compounds to their fetuses as they continue to accumulate new organochlorines in their blubber. "During lactation," wrote Beckmen and colleagues, "not only is she mobilizing her highly OC [organochlorine]-contaminated blubber, she is also transferring to her pup OCs acquired from foraging around the Pribilofs" (Beckmen, Krahn, and Stott 1997). Bioaccumulated organochlorines also have been found in many of the seals' food sources, including squid, pollack, herring, and salmon. Squid had the highest levels of dioxins among this group and salmon the lowest.

Pregnant fur seals accumulate the contaminants, passing them to their offspring in mother's milk, and thus, according to Kimberlee Beckmen and colleagues (1997), "Northern fur seal pups receive the highest doses of OC exposure *in utero* and in the neonatal period. Studies in humans and laboratory animals indicate this is a period when the immune and neurological systems are in critical developmental stages and are extremely sensitive to permanent damage from low-dose exposures of environmental contaminants."

Beckmen and colleagues (1997) wrote that "The northern fur seal (*Callorhinus ursinus*) population that breeds on St. George Island,

Alaska, has been declining at an annual rate of approximately 6 percent per year since 1973. Previous studies have failed to identify an exact cause, but have found lower-than-expected return rates after initial post-weaning migrations. Since preliminary studies suggested the need to examine the possibility of environmental contaminant-linked immunosuppression, pups and their dams were sampled during the 1996 breeding season on St. George Island" (Beckmen, Krahn, and Stott 1997).

## POPS AND ANIMALS OF THE GREAT LAKES

The Great lakes of North America have become a nerve center for study of toxic hot spots affecting animals that have been exposed to POPs. Dioxins, for example, now appear to have been responsible to some degree for the extinction of the native lake trout in the Great Lakes. Fishery officials earlier had blamed the trout's population crash during the 1950s on overfishing, habitat destruction, and predation by an introduced parasite, the sea lamprey.

Research in the Great Lakes Basin on persistent toxic substances has shown population decreases in wildlife, increased mortality rates, reproductive problems, genital malformations, birth defects such as cross-billed ducks, and tumors in fish. Impacts on human health include the presence of toxic substances in breast milk, increased incidence of cancer, and changes to the human reproductive system.

Theodora Colborn earned a doctorate in zoology from the University of Wisconsin during her fifties (in the 1980s) and got a job surveying the environmental health of the Great Lakes. Much of the visible pollution of the Great Lakes system had been dealt with by that time; the days were over when a river in Cleveland could catch fire (as it did in 1969) and create sensational headlines. By the late 1980s, many people thought the Great Lakes were being brought back to environmental health.

After two months immersing herself in reports of animal health in the Great Lakes, Colborn was convinced that the proclamations were premature.

She had come to doubt that the lakes, however improved, were truly "cleaned up." The broken eggs littering the bird colonies may have disappeared, but biologists working in the field were still reporting things that were far from normal: vanished mink populations; unhatched eggs, deformities such as crossed bills, missing eyes, and clubbed feet in cormorants, and a puzzling indifference in usually vigilant nesting birds about their incubating eggs. (Colborn, Dumanoski, and Myers 1996, 14)

Colborn was beginning to trace many reports that would later associate a wave of seal deaths in northern Europe with a large number of deaths among striped dolphins in the Mediterranean Sea. She would find the common ties between these extinctions and examinations of deformed sperm in Denmark.

At roughly the same time Colborn was suspecting that something was amiss in the Great Lakes, Niels Skakkebaek, a reproductive researcher at the University of Copenhagen, was searching for reasons why the rate of testicular cancer had risen threefold in Denmark between the 1940s and the 1980s. He was peering through microscopes at sperm samples that "might have two heads [and] . . . two tails, while another might have no head at all" (Colborn, Dumanoski, and Myers 1996, 9). Along the way, he would discover that worldwide human sperm counts seemed to be declining significantly, as much as 50 percent in half a century.

Between 1950 and 1980, the seal population collapsed near the mouth of the Rhine River. Per Reinjders, a Dutch biologist, found impaired reproductive capacity in seals living in the western reaches of the Wadden Sea, into which the Rhine River was pouring PCBs and other organochlorines. Most often, fertilized eggs failed to implant in the seal mothers' uteri.

In gull and tern colonies in the Great Lakes, the Pacific Northwest, California, and Massachusetts, field researchers have found nests with twice the usual number of eggs, which is a sign that the birds occupying the nests were two females instead of the expected male-female pair. In some Lake Ontario colonies, birds showed behavioral aberrations, including less inclination to defend their nests or sit on their eggs, which increased predation and diminished the hatching and survival of chicks.

## POPS AND ALLIGATORS IN FLORIDA'S LAKE APOPKA

POPs have been linked to stunted penises and reproductive failure in the alligators of Florida's Lake Apopka. Alligator eggs collected there had relatively high levels of a variety of contaminants, including toxaphene, dieldrin, and the DDT breakdown products DDE and DDD. Abnormally small penises were the most dramatic symptom of a profound but otherwise invisible disruption of the alligators' internal reproductive organs resulting from skewed hormone levels. These problems were later found among alligators in other Florida lakes as well, but nowhere else as intensely as at Lake Apopka, which experienced a chemical spill in its recent history. Discovery of alligator hormone abnormalities and reproductive failure in other Florida lakes indicates that chronic contamination from agricultural pesticides may be as hazardous (and on a much wider scale) than individual chemical spills.

Lake Apopka, contaminated by pesticide and herbicide runoff as well as an extensive spill of the organochlorines dicofol and DDT, suffered a progressive decline of its alligator population that began during the 1980s and continued into the 1990s. By 1995 the alligator population of the lake had declined 90 percent from its 1970s levels. Estrogen levels in young female alligators in Lake Apopka were found to be twice those of alligators in an unpolluted lake. Lake Apopka alligators also have experienced abnormal ovarian structures. Male juveniles had significantly depressed levels of testosterone (three times lower than controls), poorly organized testes, and small penises. Gonads of juvenile male alligators have been permanently altered during fetal development, disrupting production of sex hormones that govern sexual differentiation (Guillette et al. 1994).

Louis Guillette of the University of Florida, who studies alligator deformities, said that healthy alligators might have eggs that were 75 percent healthy and 25 percent dead or deformed. When scientists began to study the alligators in Lake Apopka, they found that nearly 90 percent of their eggs had died (Cadbury 1997, 124).

Many Lake Apopka alligators also were found to be "intersex," having mixed gender characteristics. An intersex male might have a penis and testes but also produce female hormones. In some cases, testes resembled ovaries. According to Guillette, "Almost 80 percent of the male alligators caught on this lake have some kind of abnormal phallus or abnormal penis" (Cadbury 1997, 126). Some of the animals changed their sex during embryonic development. The same problems seemed also to afflict other animals in Lake Apopka. Turtles, for example, were about 25 percent intersex, according to Guillette (Cadbury 1997, 126).

Alligator reproductive problems have come to light all across southern Florida (CNN 1998). In the Everglades, which are contaminated with numerous pesticides, "full-grown alligators weigh hundreds of pounds less than alligators elsewhere in Florida" (Montague 1998). In this area, juvenile alligators have levels of reproductive hormones in their blood that are far below normal. Male alligators exposed to pesticides in Florida are having difficulty reproducing, partly because their penises are not developing to normal size. Males of many wildlife species of birds, fish, amphibians, and mammals are said to have been feminized by exposure to low levels of pesticides and other industrial chemicals (Raloff 1992, 1993, 1994).

Panthers in Florida's Everglades National Park and Big Cypress swamp have suffered from sterility and sperm abnormalities, malfunctioning thyroid glands, and impaired immune response. Problems with undescended genitals in male cubs have increased markedly since 1975, a result, perhaps, of the same pollution that has been afflicting alligators in some of the same areas.

The Florida panther (*Felis concolor coryi*), an endangered species, has been failing to reproduce. By the early 1990s, only thirty to fifty Florida panthers remained. Researchers have reported that between 1985 and 1990, 67 percent of male panthers were born with one or more undescended testicles, a condition known as cryptorchidism. In England and the United States, cryptorchidism has more than doubled in men during the last four decades. Furthermore, some Florida panthers are sterile while others produced abnormal or deformed sperm (Giwercman and Skakkebaek 1992; Montague 1994). These reports surfaced at nearly the same time as reports asserting a 50 percent decline in human semen quality (Carlsen et al. 1992).

A majority of the remaining Florida panthers have exhibited several developmental abnormalities as well as reproductive defects. Twelve males had low ejaculate volumes and low sperm concentrations. A very high proportion (92.9 percent) of their sperm exhibited morphological abnormalities. Female Florida panthers had high body burdens of various contaminants, including DDE, PCBs, methoxychlor, oxychlordane, and t-nonachlor. Reproductive abnormalities in the panthers may be due to contamination of the mothers by endocrine-disrupting chemicals, according to one study (Danish Ministry of Environment and Energy 1995).

## DEATHS OF VULTURES IN INDIA AND DDT-CONTAMINATED CARRION

In some areas of India, vultures have vanished because DDT contaminated their food chain (as well as that of human beings in the area). Forty years after environmentalist Rachel Carson exposed the dangers of DDT in *Silent Spring*, large amounts of the pesticide continue to be manufactured by India's public-sector companies for health programs. The chemical is still the only practical defense against malaria in much of the tropical world. The Stockholm Convention on POPs grants DDT an extended phase-out period because of its vital role in containing malaria.

Investigations were begun during 1999 to determine why vultures had vanished from the Keoladeo National Park in western Rajasthan state, where they had been plentiful just a few years earlier. Vultures, which feed on the carcasses of DDT-contaminated dead cows, have been disappearing. DDT affects embryonic development in birds directly or by reducing eggshell thickness as well as altering mating and parenting behavior. "What is important to note is that vultures and human beings share the same food chain and must therefore be accumulating the same toxins," said Anil Agarwal, an investigator on the scene. "India's apathetic administration is yet to wake up to the pesticide threat" (Raj 2000).

India banned DDT for agricultural uses in 1998, but because it is cheap, available, and easy to use, large quantities were still being sprayed on crops two years later, according to Jagdish Prasad Singh, pesticides expert at the Voluntary Health Association of India (Raj 2000). Besides India, China and Mexico were still manufacturing DDT in 2000. In the meantime, some pests have developed resistance to DDT, as DDT levels rise in human populations where the chemical is still being used, including India. Dr. T. S. Kathpal, a pesticide chemist at the Haryana Agriculture University at Hissar, believes that virtually no item of food available in India is now free of DDT. Kathpal, who spent two decades tracking DDT contamination in animals and human beings, said Indians now show a 22.8 p.p.m. body burden of DDT compared to 4.3 for Japanese 3.70 for Australians and 2.24 for people in the United States (Raj 2000).

Greenpeace scientist Pat Costner asserted that DDT and other POP chemicals pose a particular danger to infants in South Asian countries because they interfere with lactation in mothers. "In South Asian countries, breast-feeding is not a choice but the only nutrition available to millions of newborns and the effect of POPs on lactation can be disastrous," Costner said (Raj 2000).

## TOXICITY SAVED THE WHALES?

Toxic contamination may be saving some whales from extinction because human consumers are avoiding them. The major consumers of whale meat, the Japanese, began to question its safety after high levels of heavy metals and chemicals were found in whale meat. Some Japanese scientists advised against eating whale meat, and sales declined. Japanese retailers, including one three-hundred-store supermarket chain, were reported to be removing whale meat from their shelves after scientists recommended an "immediate ban on the sale of all contaminated products" (Lean 2000). The scientists pointed out that toxic chemicals can bioaccumulate in whales and dolphins to 70,000 times the levels found in the waters in which they swim and feed.

Conservationists have been campaigning to stop whaling for more than thirty years as unregulated whaling has forced many species, such as bluefin and humpback, to the verge of extinction. Consumer-driven hunting of these and other species has continued. During 1999, however, two Japanese toxicologists and two geneticists from Harvard University analyzed more than one hundred samples of the meat served or sold in restaurants, shops, and markets across Japan in a study coordinated by the British-based Whale and Dolphin Conservation Society and the Swiss Coalition for the Protection of Whales. According to a news account in the *London Independent,* "They were astonished at the results. About half of all the samples proved to be contaminated with heavy metals or dangerous chemicals including mercury, dioxins, DDT and PCBs above maximum levels allowed for human consumption under Japanese and international standards" (Lean 2000).

During the mid- to late 1990s, a seven-year study in the Faeroe Islands found that children whose mothers had eaten contaminated whale meat during pregnancy were much more likely to suffer brain and heart damage. During 2000 researchers led by Koichi Haraguchi, associate professor at Daiichi University's College of Pharmaceutical Sciences in Fukuoka, Japan, detected dioxin levels up to 172 times the tolerable daily intake in whale and dolphin meat that was offered for sale. Haraguchi's team examined levels of dioxin as well as dibenzofuran and coplanar polychlorinated biphenyl, which have a degree of toxicity similar to that of dioxin, in thirty-eight types of whale and dolphin meat sold in Japan during 1999 and 2000. The highest concentrations of dioxins were detected in the fat of dolphins, at a maximum of 691 picograms per gram when the amount was converted into the most toxic dioxin, and 232 picograms on average, according to the report ("Whale and Dolphin Meat" 2000).

The World Wildlife Fund has warned that an initial toxic analysis of Norwegian Minke whale meat and blubber samples destined for human consumption revealed more than fifty polychlorinated biphenyl (PCB) isomers, including some dioxin-like PCBs, which are suspected endocrine disrupters (Allsopp, Costner, and Johnston 1995). The same tests also revealed twenty-five metals, including organic mercury. The Convention for International Trade of Endangered Species in Flora and Fauna (CITES) has rejected Norway's proposal to reopen international trade in whale products. Norway has been holding blubber stockpiles in the hope that one day the international ban on whale meat products may be lifted.

## ESTROGENIC CHAOS: REPORTS ON SEVERAL SPECIES

Exposure of pregnant rats to relatively low levels of dioxin did not cause maternal toxicity but did permanently reduce their male offspring's sperm counts. Testicular weight also was reduced. In another study, these effects were shown to occur after a single very low dose of dioxin (64 nanograms/kilogram) given on the fifteenth day of pregnancy. No effect on other organ systems has ever been observed at such a low dose. Mably and colleagues (1992a, 1992b) concluded that the male reproductive system "is highly sensitive to perinatal (prenatal and lactational) exposure of dioxins" (Allsopp, Costner, and Johnston 1995).

Exposure to dioxins during pregnancy causes fetal death in monkeys and other small laboratory mammals. Prenatal exposure to PCBs has been found to cause fetal malformations (teratogenic) in laboratory mammals (Shane 1989). Exposure to the organochlorine perchloroethylene (tetrachloroethylene), which is used in the dry-cleaning industry, causes fetal death in rats and growth disturbances in mice (Schwetz, Leontg, and Gehring 1975). Breeding difficulties in several species of sea fish have also been associated with exposure to organochlorines, including PCBs, DDT, and DDE (Johnston and McCrea 1992).

Chronic, long-term, low-dose exposure to dioxin in marmoset monkeys resulted in changes in lymphocytes at body burdens of only 6 to 8 nanogram per kilogram (Neubert et al. 1992). "Results from such animal experiments are of concern because effects on the immune system are occurring at body burdens similar to, or within an order of magnitude of current human body burden estimates (5–10 ng/kg), (U.S. EPA 1994). Female monkeys given a few millionths of a gram of dioxin showed a high incidence of abortions. The same monkeys also developed bone-marrow abnormalities and hemorrhages of the heart and lungs, as well as gangrene in their toes and fingers. A report on this research observed that "After 33 months the monkeys were taken off the tests but continued to deteriorate. They lost their hair, their eyelids swelled up and, finally, most of them died" (Cook and Kaufman 1982, 5).

## POPS AND BALD EAGLES

The bald eagle has lived in North America for roughly a million years as an opportunistic carnivore that usually eats carrion and steals food gathered by other birds. The bald eagle also sometimes eats small live birds and other animals. A bald eagle's preferred diet is rich in protein contaminated with fat—a perfect biological medium for bioaccumulating POPs.

During the bald eagle's million-year residency in North America, the last century has been the most threatening to its survival as a species. During the late nineteenth century, bounties were paid for dead eagles, and their customary habitat was reduced by human encroachment. By the early twentieth century, the bird had achieved a protected status as a U.S. national symbol and was recovering. During the 1940s, however, populations began to plunge again with the onset of organochlorine pollution. The travails of the eagle with DDT were chronicled in detail by Rachel Carson in *Silent Spring* (1962). By the middle 1980s, fewer than fifty bald-eagle nesting sites were known in the entire Great Lakes watershed (Colborn et al. 1990).

Bald eagles around Lake Superior get an extra dollop of organochlorine pollution because their diet is made up largely of fish-eating gulls: "Because of this taste for gulls, they accumulate concentrations of contaminants that are 20 times greater than they would have if they had eaten only fish" (Colborn, Dumanoski, and Myers 1996, 154). This level of toxicity is reflected in rising number of deformed chicks even after (as in the case of St. Lawrence River beluga whales) human-induced pollution had ceased or been reduced significantly.

Rachel Carson (1962) described how organochloric chemicals cause eggshell thinning, increased embryo mortality and deformities, and caused feminization of males in many bird species. Carson's case centered on bald eagles, whose populations dropped drastically during the 1950s and 1960s. The poisoning of avian species that Carson described so graphically has continued and intensified in some areas (note India's dying vultures, above). Studies of fish-eating bird species around the Great Lakes have shown that eggs contain organochlorine pesticides, PCBs, and dioxins (PCDDs and PCDFs) in concentrations high enough to cause adverse effects on parents and their chicks. Bird populations in this area have been declining since the 1950s (Giesy, Ludwig, and Tillitt 1994).

Gulls in Puget Sound and the Great Lakes have shown signs of eggshell thinning and abnormalities of their reproductive tracts, including feminization of male embryos. Populations have declined in some areas, and sex ratios have been skewed. These areas are contaminated with residues of DDT, PCBs, and polycyclic aromatic hydrocarbons that may be responsible for these effects singly or in combination (Fry 1995; Schettler et al. 1999, 156–58).

Despite restrictions on the use of DDT in the United States after 1972, embryonic and chick survival along the shore of the Great Lakes continued, at least until about 1990, to be inadequate to maintain stable populations among several bird species (Colborn, vom Saal, and Soto 1993). Newly hatched male chicks from Lake Ontario had testes resembling ovaries. In addition, female pairs of gulls were found tending abnormally large clutches of eggs, most of which were infertile.

Michael Fry of University of California at Davis has been observing "lesbian gulls," female-female pairs on Santa Barbara Island off the California coast. Male gulls seem, according to Fry's observations, to have lost interest in mating and sex. Many male gulls display feminized sex organs after they have been chemically castrated by DDT and other environmental pollutants (Montague 1994).

## ESTROGEN MIMICS AND ZEBRA FINCHES

Estrogen mimics have been implicated with masculinizing female zebra finches and causing infertility in males of the same species. The research indicates that zebra finches' consumption of estrogen-mimicking chemicals can alter the brains of the female birds enough to make them sing. In nature, male finches sing to attract mates. Scientists who fed female zebra finch hatchlings for a week with estradiol, an estrogen hormone used in hormone replacement therapy, produced singing birds ("Environmental Estrogens" 2002). Millions of women take estrogen supplements in birth-control pills and in estrogen supplements to relieve hot flashes and other symptoms of menopause. These pharmaceutical estrogens have been detected in some cities' treated sewage wastewater.

Dissecting the birds' brains, the scientists discovered that the regions controlling singing were well-developed in the females treated with estradiol. The control group, given canola oil instead, showed no such effect. Estradiol-treated male chicks were found to be infertile as adults ("Environmental Estrogens" 2002). "These results indicate that songbird populations may be at risk if they are exposed to estrogenic chemicals as chicks," said James Millam, professor in the animal science department in the University of California at Davis College of Agricultural and Environmental Sciences and lead author of the studies published in the December 2001 and April 2002 issues of the journal *Hormones and Behavior* ("Environmental Estrogens" 2002).

In an earlier study, Millam and colleagues found that estradiol hindered the finches' ability to reproduce. According to a report by the Environment News Service, "Exposure to estradiol resulted in fewer and more brittle eggs for the tested group and more infertile eggs in groups containing estrogen treated males. As a result, the number of hatchlings decreased compared to those of the control group" ("Environmental Estrogens" 2002).

## HOW CHEMICAL PESTICIDES SOMETIMES INCREASE INSECT INFESTATIONS: THE ARMYWORM

Malathion has been used to eradicate the boll weevil in the cotton fields of the South, but in so doing it also nearly eradicated a wasp that eats armyworms. As a result, a plague of armyworms in recent years has driven some cotton farmers into bankruptcy. "It's unbelievable," said Auburn University ornithologist Geoffrey E. Hill. "I've literally seen them strip a field. There's nothing left" (Williams 1999, 26). Philip Barbout, a Mississippi farmer, said that "In 1995 there were so many, they were crawling up telephone poles" (Williams 1999, 26).

American Cyanamid has been requesting EPA approval to use Pirate, a new chemical, one of a family called pyrroles. Pirate was being used in about thirty other countries (including Australia, China, South Africa, Zambia, and Zimbabwe) by the year 2000. Some environmentalists have warned that chlorfenapyr, the active ingredient in Pirate, may be the next DDT. Like other organochlorines, chlorfenapyr accumulates in animals' body fat and disrupts their endocrine and immune systems. The American Bird Conservancy issued an action alert, requesting popular pressure on the EPA to deny permission to sell and use Pirate. An EPA risk assessment has said that residues of the chemical "present a substantial risk to avian species for both acute lethal effects and impairment of reproduction" (Williams 1999, 26). American Cyanamid already has obtained emergency permits to use the chemical in several states, despite the fact that a laboratory study ordered by the EPA showed that mallard ducks exposed to a level of the chemical said to be typical of the wild laid 40 percent fewer eggs (which had a lower hatching rate and birth rate) than a control group.

## REFERENCES

Allsopp M., Pat Costner, and Paul Johnston. 1995. *Body of Evidence: The Effects of Chlorine on Human Health*. London: Greenpeace International. http://www.greenpeace.org/~toxics/reports/recipe.html.

Allsopp, Michelle, Ben Erry, Ruth Stringer, Paul Johnston, and David Santillo. 2000. "Recipe for Disaster: A Review of Persistent Organic Pollutants in Food." Exeter, U.K.: Greenpeace Research Laboratories and Department of Biology University of Exeter. http://www.green peace.org/~toxics/reports/recipe.html.

Arnold, D. L., J. Mes, and F. Bryce. 1990. "A Pilot Study on the Effects of Aroclor 1254 Ingestion by Rhesus and *Cynomolgus* Monkeys As a Model for Human Ingestion of PCBs." *Food Chemistry and Toxicology* 28: 847–57.

Barsotti, D. A., R. J. Marlar, and J. R. Allen. 1976. "Reproductive Dysfunction in Rhesus Monkeys Exposed to Low Levels of Polychlorinated Biphenyls (Aroclor 1248)." *Food Chemistry and Toxicology* 14: 99–103.

Beckmen, Kimberlee B., Margaret M. Krahn, and Jeffrey L. Stott. 1997. "Immunotoxicology of Organochlorine Contaminants in Northern Fur Seals, *Callorhinus Ursinus*." Past Arctic Research Initiative (ARI) Progress Reports: 1997. http://www.cifar.uaf.edu/proposal/award97/award97.html.

Brodkin, Marc, and Martin Simon. 1997. "The Effects of Aquatic Acidification on Rana Pipiens." *Froglog* 20 (January): 3. http://acs-info.open.ac.uk/info/newsletters/FROGLOG.html.

Cable News Network (CNN) Interactive. 1998. "Pesticides Suspected in Florida Gator Decline." March 15. http://www.cnn.com/EARTH/9803/15/gator.woes/index.html.

Cadbury, Deborah. 1997. *Altering Eden: The Feminization of Nature*. New York: St. Martin's Press.

Campbell, J. S., J. Wong, L. Wong, D. Tryphonas, E. Arnold, B. Nera, B. Cross, and B. LaBossiere. 1985. "Is Simian Endometriosis an Effect of Immunotoxicity?" Paper presented at the Ontario Association of Pathologists 48th Annual Meeting, London, Ontario.

Carlsen E., A. Giwercman, N. Keiding, and N. E. Skakkebaek. 1992. "Evidence for Decreasing Quality of Semen during the Past 50 Years." *British Medical Journal* 305: 609–13.

Carson, Rachel. 1962. *Silent Spring*. Boston: Houghton-Mifflin.

Chow, Gee. 1997. "Pesticides and the Mystery of Deformed Frogs." *Journal of Pesticide Reform* 17 (Fall): 14.

Colborn T., A. Davidson, S. N. Green, R. A. Hodge, C. I. Jackson, and R. A. Liroff. 1990. *Great Lakes/Great Legacy?* Washington, D.C.: Conservation Foundation.

Colborn, T., D. Dumanoski, and J. P. Myers. 1996. *Our Stolen Future: Are We Threatening Our Fertility, Intelligence, and Survival? A Scientific Detective Story*. New York: Penguin.

Colborn, Theo, Frederick S. vom Saal, and Ana M. Soto. 1993. "Developmental Effects of Endocrine-Disrupting Chemicals in Wildlife and Humans." *Environmental Health Perspectives* 101, no. 5: 378–84.

Cook, Judith, and Chris Kaufman. 1982. *Portrait of a Poison: The 2,4,5-T Story*. London: Pluto Press.

Dalton, Rex. 2001. "Frogs Put in the Gender Blender by America's Favourite Herbicide." *Nature* 416 (April 18): 665–66.

Danish Ministry of Environment and Energy. 1995. *Male Reproductive Health and Environmental Chemicals with Oestrogenic Effects*. Copenhagen: Danish Environmental Protection Agency.

Dicke, William. 1996. "Numerous U.S. Plant and Freshwater Species Found in Peril." *New York Times*, January 2.

"Environmental Estrogens Could Hamper Songbird Breeding." 2002. Environment News Service. May 29. http://ens-news.com/ens/may2002/2002-05-29-09.asp.

European Commission. 1997. *European Workshop on the Impact of Endocrine Disrupters on Human Health and Wildlife: Report of Proceedings, December 2–4, 1996, Weybridge, U.K.* Copenhagen, Denmark: European Commission. [Report EUR 17549.]

Folmar, Leroy C., et al. 1996. "Vitellogenin Induction and Reduced Serum Testosterone Concentrations in Feral Male Carp (*Cyprinus Carpio*) Captured near a Major Metropolitan Sewage Treatment Plant." *Environmental Health Perspectives* 104: 1096–101.

Forberg, S., O. Tjelvar, and M. Olsson. 1991. "Radiocesium in Muscle Tissue of Reindeer and Pike from Northern Sweden before and after the Chernobyl Accident: A Retrospective Study on Tissue Samples from the Swedish Environmental Specimen Bank." *Science of the Total Environment* 115: 179–89.

Fry, D. M. 1995. "Reproductive Effects in Birds Exposed to Pesticides and Industrial Chemicals." *Environmental Health Perspectives* 103, suppl. 7: 165–71.

Giesy, J. P., J. Ludwig, and D. E. Tillitt. 1994. "Deformities in Birds of the Great Lakes Region." *Environmental Science and Technology* 28, no. 3: 128–35.

Giwercman, A., and N. E. Skakkebaek. 1992. "The Human Testis: An Organ at Risk?" *International Journal of Andrology* 15: 373–75.

Gray, L. E., W. R. Kelce, E. Monosson, J. S. Ostby, and L. S. Birnbaum. 1995. "Exposure to TCDD during Development Permanently Alters Reproductive Function in Male Long Evans Rats and Hamsters: Reduced Ejaculated and Epididymal Sperm Numbers and Sex Accessory Gland Weights in Offspring with Normal Androgenic Status." *Toxicology and Applied Pharmacology* 131: 108–18.

Green, Emily. 2002. "Common Weed Killer Causes Sexual Abnormalities in Frogs, Study Claims." *Los Angeles Times,* April 16.

Guillette, L. J., T. S. Gross, G. R. Masson, J. M. Matter, H. F. Percival, and A. R. Woodward. 1994. "Developmental Abnormalities of the Gonad and Abnormal Sex Hormone Concentrations in Juvenile Alligators from Contaminated and Control Lakes in Florida." *Environmental Health Perspectives* 102, no. 9: 680–88.

Halliday, Tim. 1997. "1996 International Union for Conservation of Nature Red List." *Froglog* 21 (March): 2. http://acs-info.open.ac.uk/info/news letters/FROGLOG.html.

Hayes, Tyrone B., Atif Collins, Melissa Lee, Magdelena Mendoza, Nigel Noriega, A. Ali Stuart, and Aaron Vonk. 2002. "Hermaphroditic, Demasculinized Frogs after Exposure to the Herbicide Atrazine at Low Ecologically Relevant Doses." *Proceedings of the National Academy of Sciences* 99, no. 8 (April 16): 5476–80.

"Heavy Metal Levels in Reindeer, Caribou, and Plants of the Seward Peninsula." 2000. Report, Reindeer Research Program, University

of Alaska at Fairbanks, April. http://reindeer.salrm.alaska.edu/
research.htm.

Hong, R, K. Taylor, and R. Abanour. 1989. "Immune Abnormalities Associated with Chronic TCDD Exposure in Rhesus." *Chemosphere* 18:
313–20.

Johnston, P., and I. McCrea, eds. 1992. *Death in Small Doses.* London:
Greenpeace International.

Kirby, Alex. 2000. "Scientists Test Sex-Change Bears." British Broadcasting
Company News. September 1. http://irptc.unep.ch/pops/newlayout/
press_items.htm.

Lean, Geoffrey. 2000. "Poison Saves Hunted Whales." *London Independent,*
January 9.

"Links Found to Frog Deformities." 2002. Associated Press, July 8. http://
www.cnn.com/2002/US/07/08/deformed.frogs.ap/index.html.

Lopez-Martin, J. M., J. Ruiz-Olmo, and S. P. Minano. 1994. "Organochlorine Residue Levels in the European Mink (*Mustela Lutreola*) in Northern Spain." *Ambio* 3, nos. 4–5: 294–95.

Mably, Thomas A., R. W. Moore, D. L. Bjerke, and R. E. Peterson. 1991.
"The Male Reproductive System Is Highly Sensitive to *in Utero* and
Lactational TCDD Exposure." *Banbury Reports* 5: 69–78.

Mably, Thomas A., D. L. Bjerke, R. W. Moore, A. Gendron-Fitzpatrick, and
R. E. Peterson. 1992a. "*In Utero* and Lactational Exposure of Male
Rats to 2,3,7,8-Tetrachlorodibenzo-P-dioxin. 1. Effects on Androgenic Status." *Toxicology and Applied Pharmacology* 114 (May): 97–
107.

———. 1992b. "*In Utero* and Lactational Exposure of Male Rats to 2,3,7,8-
tetrachlorodibenzo-p-dioxin. 2. Effects on Spermatogenesis and Reproductive Capability." *Toxicology and Applied Pharmacology* 114:
118–26.

Montague, Peter. 1993. "How We Got Here, Part 1; The History of Chlorinated Biphenyl (PCBs)." *Rachel's Environment and Health News* 327,
March 4. http://www.rachel.org/bulletin/bulletin.cfm?Issue_ID = 802
&bulletin_ID = 48.

———. 1994. "PCBs Diminish Penis Size." *Rachel's Environment and Health
News* 372, January 13. http://www.rachel.org/bulletin/index.cfm?
St = 2.

———. 1997a. "Fish Sex Hormones." *Rachel's Environment and Health
News* 545, May 8. http://www.rachel.org/bulletin/index.cfm?St = 2.

———. 1997b. "The Weybridge Report." *Rachel's Environment and Health
News* 547, May 22. http://www.rachel.org/bulletin/index.cfm?St = 2.

———. 1998. "Frogs, Alligators, and Pesticides." *Rachel's Environment and
Health News* 590, March 19. http://www.rachel.org/bulletin/index.
cfm?St = 2.

Nelson, Bryn. 2002. "Frogs Feel Effect of a Herbicide; Sexual Damage Includes Loss of Voice in Males." *Newsday,* April 16.

Netting, Jessa. 2000. "Pesticides Implicated in Declining Frog Numbers." *Nature* 408 (December 4): 760.

Neubert, R., G. Golor, R. Stahlman, H. Helge, and D. Neubert. 1992. "Polyhalogenated Dibenzo-p-dioxins and Dibenzo-furans and the Immune System. 4. Effects of Multiple-Dose Treatment with 2,3,7,8-tetrachlorodibenzo-p-dioxin (TCDD) on Peripheral Lymphocyte Subpopulations of a Non-human Primate (*Calloithrix Jacchus*)." *Archives of Toxicology* 66: 250–71.

Norstrom, R. J., and D. C. G. Muir. 1994. "Chlorinated Hydrocarbon Contaminants in Arctic Marine Mammals." *Science of the Total Environment* 154: 107–28.

Ouellet, Martin, et al. 1997. "Hind-limb Deformities (Ectromelia, Ectrodactyly) in Free-Living Anurans from Agricultural Habitats." *Journal of Wildlife Diseases* 33: 95–104.

Persistent Organic Pollutants: Background. Lindane Education and Research Network, National Pollution Prevention Roundtable. No date. http://www.lindane.org/lindane/persistant_toxic_substances.htm.

Purdom, C., et al. 1994. "Estrogenic Effects of Effluents from Sewage Treatment Works." *Chemistry and Ecology* 8: 275–85.

Raj, Ranjit Dev. 2000. "Vanishing Vultures Bode Ill for Indians." Interpress Service, January 3. http://irptc.unep.ch/pops/newlayout/press_items.htm.

Raloff, Janet. 1992. "Perinatal Dioxin Feminizes Male Rats." *Science News* 141 (May 30): 359.

———. 1993. "EcoCancers." *Science News* 144 (July 3): 10–13.

———. 1994. "The Gender Benders." *Science News* 145 (January 8): 24–27.

Rier, S. E., D. C. Martin, R. E. Bowman, W. P. Dmowski, and J. L. Becker. 1993. "Endometriosis in Rhesus Monkeys (*Macaca Mulatta*) Following Chronic Exposure to 2,3,7,8-tetrachlordibenzo-p-dioxin. *Fundamentals of Applied Toxicology* 21: 433–41.

Russell, Ronald W., and Stephen J. Hecnar. 1996. "The Ghosts of Pesticides Past?" *Froglog* 19 (November). http://acs-info.open.ac.uk/info/news letters/FROGLOG.html.

Russell, Ronald W., et al. 1995. "Organochlorine Pesticide Residues in Southern Ontario Spring Peepers." *Environmental Toxicology and Chemistry* 14: 815–17.

Russell, Ronald W., et al. 1997. "Polychlorinated Biphenyls and Chlorinated Pesticides in Southern Ontario, Canada, Green Frogs." *Environmental Toxicology and Chemistry* 16: 2258–63.

Schettler, Ted, Gina Solomon, Maria Valenti, and Anne Huddle. 1999. *Generations at Risk: Reproductive Health and the Environment.* Cambridge, Mass.: MIT Press.

Schwetz, B. A., B. M. J. Leontg, and B. J. Gehring. 1975. "The Effect of Maternally Inhaled Trichloroethylene, Perchloroethylene, Methyl Chloroform and Methyl Chloride on Embryonal and Fetal Development in Mice and Rats." *Toxicology and Applied Pharmacology* 32: 84–96.

Senanayake, Ranil, et al. 1997. "Frog Tea?" *Froglog* 23 (August): 2. http://acs-info.open.ac.uk/info/newsletters/FROGLOG.html.

Shane, B. S. 1989. "Human Reproductive Hazards." *Environmental Science and Technology* 23, no. 10: 1187–95.

Sharpe, R. M., and N. E. Skakkebaek. 1993. "Are Oestrogens Involved in Falling Sperm Counts and Disorders of the Male Reproductive Tract?" *Lancet* 341: 1392–95.

Skogland, T. 1987. "Radiocesium Concentrations in Wild Reindeer at Dovrefjell, Norway." *Rangifer* 7: 42–45.

Souder, William. 2002. "A Pesticide-Parasite Role in Frogs' Deformities?" *Washington Post,* July 15, A-9.

Sumpter, John P. 1995. "Feminized Responses in Fish to Environmental Estrogens." *Toxicology Letters* 82–83 (December): 737–42.

Sumpter, John P., and Susan Jobling. 1995. "Vitellogenesis As a Biomarker for Estrogenic Contamination of the Aquatic Environment." *Environmental Health Perspectives* 103, suppl. 7 (October): 173–77.

Thornton, Joe. 1997. "PVC: The Poison Plastic." Washington, D.C.: Greenpeace USA. http://www.greenpeace.org/~usa/reports/toxics/PVC/cradle/dcgsum.html.

Tyler, Michael J. 1997. "Herbicides Kill Frogs," *Froglog* 21 (March): 2. http://acs-info.open.ac.uk/info/newsletters/FROGLOG.html.

U.S. Environmental Protection Agency. 1994. *Health Assessment Document for 2,3,7,8-tetrachlorodibenzo-p-dioxin (TCDD) and Related Compounds.* 3 vols. Washington, D.C.: U.S. EPA, Office of Research and Development.

———. 2002. "Persistent, Bioaccumulative, and Toxics Initiative." Fact Sheet. http://www.epa.gov/opptintr/pbt.

"U.S. EPA Takes Action to Clean Great Lakes of Toxicity." 2000. Environmental Protection Agency. November 6. http://www.planetark.org/dailynewsstory.cfm?newsid = 8799.

"UV Radiation Linked to Deformed Amphibians." 2002. Environment News Service. June 21. http://ens-news.com/ens/jun2002/2002-06-21-09.asp#anchor4.

"Whale and Dolphin Meat Sold in Japan Has High Levels of Dioxin." 2000. *Japanese Times Online,* July 4. http://www.japantimes.co.jp/cgi bin/getarticle.pl5?nn20000704a4.htm.

Williams, Wendy. 1999. "Pirate Fear." *Scientific American,* October, 26.

World Wildlife Fund. 2000. "Toxics—What's New." http://www.worldwild
life.org/toxics/progareas/pop/pop_rep.htm.

Yang, Y. G., H. Lebrec, and G. R. Burleson. 1994. "Effect of 2,3,7,8-
tetrachlorodibenzo-p-dioxin (TCDD) on Pulmonary Influenza Virus
Titer and Natural Killer Activity in Rats." *Toxicology and Applied Pharmacology* 23: 125–31.

# 6

# End of the Line: The Dirty Dozen
# and Human Health

Barbara Alice Mann, a friend and colleague, recalls:

I breast-fed Tassia [her daughter], which now makes me wonder whether her allergies (and maybe, her diabetes!) are owing to that little act of mother-love. Some of our health problems relate directly to our lack of certain antigens in our blood (a Native American legacy, I hear), but the really weird allergies we both have came from indiscriminate and irresponsible chemical use by government and industry, I believe. I clearly remember walking home from grade school—I was seven or so—while the city was spraying for pests. I'm sure it was DDT. The truck came by, spewing out clouds of unbreathable toxins, covering us little kids walking along the sidewalk, our papers and books in hand. I remember coughing my way through the fog, indignant because the guys operating the truck were laughing at having caught all us little kids by surprise. However, I also recall an "infomercial" I saw later that was being run on television concerning the spraying, at a time when the nightly news was still revered as a truth-teller. The announcer said that the spraying was safe for humans, and even showed people walking through sprays with smiles on their faces. I now realize that this was the basest sort of propaganda. I wonder about the extreme asthma I have now. At seven years of age, however, I accepted the smiling faces in the chemical fog as the gospel truth. (Mann, personal communication, December 9, 2000)

Mann, who lives in Toledo, Ohio, grew up a few years before Rachel Carson's *Silent Spring* (1962) first attacked "the myth of the harmlessness of DDT" (21). Four decades later, each human denizen of the Earth carries in his or her body residues of several hundred dozen (or more) chemicals that were not present in the bodies of our great-grandparents at the end of the nineteenth century. Because everyone now carries some of these chemicals, no unexposed population exists to study as a control group.

Every child born today has been exposed to persistent organic pollutants in the womb. Because these chemicals also become concentrated in breast milk because of their affinity for fatty substances, a baby may experience his or her most intense exposure to POPs through breast-feeding. In nature, the breast-feeding of mammals is merely one more step up a food chain that bioaccumulates POPs.

Chemicals that were developed to control disease, increase food production, and improve our standard of living have, in fact, become a threat to biodiversity and human health. According to the World Wildlife Fund, "This exposure threatens the integrity of the next generation. Given these immense stakes, precaution dictates swift and strong action to eliminate the use and production of persistent chemicals. POPs by their nature cannot be managed" (World Wildlife Fund 2000).

The toxic overload of POPs on the environment is not a small matter. The National Environmental Trust, Physicians for Social Responsibility, and the Learning Disabilities Association of America have estimated that releases into the environment of developmental and neurological toxins amount to about 24 billion pounds a year— enough toxic chemicals to fill a line of railroad tanker cars stretching from New York City to Albuquerque, New Mexico. In many places, old stockpiles of pesticides and industrial chemicals are an increasing hazard to those who live nearby as they leak, leach, and evaporate into the air from dumpsites and deteriorating storage containers. Until effective and adequately funded disposal programs are put into place, new organochlorine emissions will continue to escape and add to the existing danger.

The health effects of these chemicals on human beings are significant. Recent findings in human populations indicate that certain populations are most at risk, notably people who routinely consume PCBs in fish and wildlife. According to one report, "These susceptible populations include sport anglers, Native Americans, women of child-bearing age, pregnant women, fetuses and nursing infants of mothers who consume contaminated Great Lakes fish, infants and children, the elderly, and the urban poor" (U.S. EPA 1996).

Neurobehavioral and developmental deficits occur in newborns and continue through school-age children who have experienced *in utero* exposure to PCBs. Other systemic effects, including liver disease and diabetes as well as compromises of the immune system, may be associated with elevated serum levels of PCBs. Increased cancer risks also have been associated with PCB exposures.

According to a study by the U.S. Environmental Protection Agency,

Developmental effects [of PCB exposure] included a statistically significant decrease in gestational age (by 4.9 days), birth weight (by 160 to 190 grams), and head circumference (by 0.6 centimeters). Five months post-term these effects were still evident compared to the control population. Neurobehavioral deficits observed include depressed responsiveness throughout the course of the study, impaired visual recognition, and poor short-term memory at seven months of age. At four years following birth, these deficits in weight gain, depressed responsiveness, and reduced performance on the visual recognition-memory test, one of the best validated tests for the assessment of human cognitive function, were still evident. (U.S. EPA 1996)

According to the same study, "Adult men, women beyond reproductive years, and the elderly are at an increased risk of cancer, and may also be at an increased risk of immune and endocrine system effects, from exposure to PCBs in fish."

Boiled down to a small fistful of everyday English, these findings send an electrifying message that everyone—even the cabinet members of the George W. Bush presidency, who have pledged their support of the Stockholm Convention on POPs—can understand: Stop the diffusion of POPs into the environment, and stop it soon, or future generations face a wave of cancers and reproductive disorders.

Several occupational or epidemiologic studies have indicated or demonstrated other adverse health effects related to PCB exposure, including cardiovascular, hepatic, immune, and musculo-skeletal problems, as well as several types of cancers. Kreiss, Zack, and Kimbrough (1981) reported a 30 percent increase over the national average in the incidence of hypertension in a PCB-exposed population at Triana, Alabama. Kreiss and colleagues' study associated increased serum PCB levels with increased systolic and diastolic blood pressure. Subjects' PCB levels were elevated mainly because they ate fish contaminated with the chemical.

Findings related to POPs have called into question the assumption of pharmacology that dosage levels are the most important determinant of toxicity. With some POPs, a low dose can induce more of an impact than a high dose, a situation that scientists call a non-monotonic response.

Researchers have linked organochlorine exposure to a wide variety of effects at different levels of exposure, including reproductive disorders (see chapter 7), malfunction of the nervous system, diabetes, suppression of the immune system, disruption of the endocrine system (thymus, thyroid, ovaries, testes, etc.), and reproductive and developmental disorders. Specific conditions associated with PCB exposure include reduced fertility; intellectual and attentional deficits; increased susceptibility to bacterial, viral, parasitic, and neoplastic disease; diminished intellectual, emotional, and physical capabilities associated with hormonal imbalances; and irreversible abnormalities of the brain, immune system, and reproductive organs of offspring from pre- and postnatal exposures. Average concentrations of dioxin now found among the populations of the United States and Europe are at or near those levels associated with lower levels of testosterone, the hormone that influences male sexual characteristics and the libido (sexual drive) in both men and women; altered glucose tolerance, a symptom of diabetes; and changes in the immune system (Allsopp, Costner, and Johnston 1995).

## TOP OF THE FOOD CHAIN: THE PERILS OF FISH CONSUMPTION

Bioaccumulation makes PCBs in fish especially hazardous to humans. Some PCBs persist in the body and remain biologically active long after actual exposure ceases. In addition, a U.S. Environmental Protection Agency report (1996) says, "Bioaccumulated PCBs appear to be more toxic than commercial PCBs and appear to be more persistent in the body. For exposure through the food chain [therefore], risks can be higher than for other exposures."

Some of the earliest investigations of PCBs' effects on human populations were conducted by Harold Humphrey of the Michigan Department of Public Health and his colleagues. Their work, during the 1970s and 1980s, demonstrated a correlation between levels of PCBs in fetal tissues and maternal consumption of contaminated fish (Humphrey 1983). The Michigan Maternal Infant Cohort Study (Fein, Jacobson, and Jacobson 1984; Jacobson, Jacobson, and Humphrey 1990; Jacobson, Fein, and Jacobson 1985) reported both developmental disorders and cognitive deficits in the offspring of mothers who ate contaminated fish six months prior to and during pregnancy.

Schantz and colleagues (1996) studied sports anglers fifty to ninety years of age, in two groups: high-volume fish eaters who consumed twenty-four or more pounds of fish they caught in the Great Lakes annually for more than fifteen years, and those who consumed less than six pounds annually. This study demonstrated that median levels of total PCBs, DDE, and mercury were significantly higher in high-volume fish eaters than in low-volume fish eaters. "The median serum total PCB concentration for high fish eaters was 12 p.p.b. and 5 p.p.b. for low fisheaters; the maximum values were 75 p.p.b. and 26 p.p.b., respectively. The median serum DDE concentration for high fish eaters was 10 p.p.b. and 5 p.p.b. for low fish eaters; maximum values were 145 p.p.b. and 33 p.p.b., respectively. . . . High-volume fish eaters [also] presented disproportionately higher body burden levels of PCBs and DDE than low-volume fish eaters in each decadal age group (i.e., 50–59, 60–69)" (U.S. EPA 1996). Lonky and colleagues (1996) investigated newborns' exposure to Great Lakes contaminants. Women in the high fish-consumption group reported eating an average of 388 PCB-equivalent pounds of Lake Ontario fish over sixteen years, which is equivalent to about two pounds of salmon or lake trout per month. Waller and colleagues (1996) conducted a similar investigation with African-American women and their newborns.

Dellinger, Kmiecek, and Gerstenberger (1995) studied contaminants in the diets of Ojibwa who consume Great Lakes fish. Their studies have associated elevated PCB serum concentrations with self-reported diabetes and liver disease. This correlation confirms similar findings of these investigators in the Red Cliff Band of Lake Superior Ojibwa (between 1990 and 1993) (Dellinger, Kmiecek, and Gerstenberger 1995).

Fitzgerald and colleagues (1996) studied a population of Native Americans to investigate associations between consumption of locally caught fish and wildlife and body burdens of sixty-eight types of PCBs, dichlorodiphenyl-dichloroethylene (DDE), mirex, and hexachlorobenzene (HCB). Preliminary data from this study indicated that Native American men who ate at least eight fish meals per month for at least two years before participating in the study had a mean PCB concentration of 5.4 ppb, which is higher than a general population level of 1 to 2 ppb. Serum PCB level increased with the number of fish consumed. The chemicals in question all accumulated with age. Consumers of fish retained most of their PCB burden, adding to it with each meal consumed.


## POPS AND THE IMMUNE SYSTEM

Some organochlorine chemicals are toxic to the immune system, reducing resistance to infection and tumors (Thomas 1990). Animal studies indicate that many organochlorine compounds, including PCBs, DDT, dieldrin, dioxins, hexachlorocyclohexane, hexachloro-benzene, chlordane, and pentachlorophenol, are toxic to the im-mune system (Safe 1994, 1995; Thomas 1990).

Studies of PCB exposure (Harada 1976; Hsu et al. 1985; Wong and Huang 1981) via contaminated rice oil in Japan during 1968 and Taiwan in 1979 also contribute to the overall weight of evidence that xenobiotic agents disrupt normal endocrine function and are associated with neurobehavioral deficits. These incidents were re-ferred to as yusho disease in the case of the Japanese studies and yu-cheng disease in Taiwan.

Epidemiology studies following the yusho and yu-cheng incidents in Japan and Taiwan, during which people ate rice contaminated with PCBs and dioxins, showed significant reductions in some types of immune system cells and an increase in respiratory bronchitis and skin infections in exposed individuals. Similar results were ob-tained in studies following a dioxin release in Times Beach, Mis-souri, during 1983 that forced the abandonment of the town (Allsopp, Costner, and Johnston 1995).

Following the PCB rice-oil contamination in Japan and Taiwan, infants born of exposed mothers exhibited a range of neurobehav-ioral problems. Cognitive testing "showed significantly lower overall age-adjusted developmental scores in the exposed children" (U.S. EPA 1996). Delays were observed at all age levels; they were greater in children who were "smaller in size, had neonatal signs of intoxi-cation and/or had a history of nail deformities" (U.S. EPA 1996). Follow-up intelligence testing when the children were four to seven years of age indicated that effects on cognitive development per-sisted for several years following exposure (Chen, Guo, and Hsu 1992). These studies indicated that the presence of dibenzofurans in contaminated rice oil (with PCBs) may have made the symptoms worse. Furans were produced when the rice oil was heated prior to human consumption.

## POPS' EFFECTS AND HIV/AIDS

Studies published during the 1990s raised a possibility that POPs may act to suppress the immune system in ways similar to those of the HIV/AIDS virus. The synergism of the human immunodeficiency virus 1 (HIV-1) and dioxin on the immune system may be more deleterious than either by itself; both appear to have similar targets in the immune system, causing suppression of certain lymphocyte cells (T-cells) (Allsopp, Costner, and Johnston 1995). Comments Joe Thornton, "The recent finding that the genome of the HIV-1 virus contains regulatory sequences that bind the dioxin-receptor complex and activate transcription of viral genes is cause for concern that dioxin-like chemicals may also play a role in the expression of infectious disease" (Yao et al. 1995, 366).

Dioxins, PCBs, and a number of other POPs undermine the immune system by interfering with antibodies. Some organochlorines (including PVC, PCBs, and some chlorinated solvents) "have been associated with the development of autoimmunity, in which the immune system attacks the body's own tissues as foreign" (Thornton 2000, 62). Autoimmune diseases include lupus, rheumatoid arthritis, and systemic sclerosis.

Human studies in Sweden and Canada have linked dietary intake of PCBs and other POPs to immune system dysfunction. The Swedish study noted a correlation between the level of PCBs, dioxins, and furans in a given diet and significant reductions in the populations of natural killer cells that play a key role in the body's defenses against cancer. Canadian researchers reported that children who were exposed to high levels of POPs experienced rates of infection ten to fifteen times higher than those of comparable, unexposed children. A recent Dutch study exploring background levels of contaminants on children's development linked immune system changes in infants to their exposure to PCBs and dioxins before and around the time of birth. The researchers noted that these problems may presage such later difficulties as immune suppression, allergies, and autoimmune diseases (World Wildlife Fund 2000).

In North Carolina, a group of infants was followed from birth to examine the effects on the offspring of mothers with "normal" levels (about 1 to 2 ppm) of PCBs in their bodies. Among 866 North Carolina infants who were tested, higher PCBs in mother's milk "was correlated with hypotonicity (loss of muscle tone) and abnormally weak reflexes. Subsequent studies of 802 North Carolina children at ages six months and twelve months revealed those with higher levels of PCBs had poorer performance on tests requiring fine motor coordination" (Montague 1994). Researchers reviewing the history of these children concluded, "There is thus consistent evidence that prenatal exposure to levels of PCBs commonly encountered in the U.S. produces detectable effects on motor maturation and some evidence of impaired infant learning" (Tilson et al. 1990, 239).

## ARE CANCER RATES RISING?

A National Cancer Institute report issued during 2000 states that the incidence rate (the number of new cancer cases per 100,000 persons per year) for all cancers declined an average of 0.8 percent per year between 1990 and 1997. "The greatest decrease in incidence (at a rate of 1.3 percent per year) occurred after 1992," according to this report (Ries et al. 2000, 2398). This is not to imply, says the report, that environmental carcinogens do not exist; mortality rates also are down as a result of earlier detection and better cancer treatments, making the level of cancer incidence a difficult proxy for contamination of the environment.

The same report also concludes that "Breast cancer incidence rates have shown little change in the 1990s, while breast cancer death rates have declined about 2 percent per year since 1990 and have dropped sharply since 1995" (Ries et al. 2000, 2398). "Incidence and death rate for non-Hodgkins lymphoma among women are continuing to increase, while incidence rates for melanoma for both sexes combined have continued to rise about 3 percent annually since 1981, [although] death rates have been approximately level since 1989" (Ries et al. 2000, 2398).

Declines in general cancer rates may be masking, to some extent, concurrent rises in cancers attributable to POP exposure. Rates of some cancers have risen markedly. For example, according to Phil Landrigan, a pediatrician and chairman of preventative medicine at Mount Sinai School of Medicine, since 1972, the date at which national records began in the United States, the rate of brain cancer incidence among children has increased 41 percent (Moyers 2001).

As Robert Napier commented in London's *Guardian:*

Currently one person in three will get cancer and this figure will rise. The idea that cancer is due to poor lifestyle, bad genes, or viruses is being increasingly discredited. The massive increase in cancer in industrialized nations is partially due to the release of 100,000 synthetic chemicals into the environment, their concentration in the food chain, and their bioaccumulation in humans. Each of us carries between 300 and 500 man-made chemicals in our body. (Napier 2001, 19)

A seven-year study of breast cancer clusters on Long Island, concluded during 2002, found no link with organochlorine compounds. The Long Island Breast Cancer Study Project was one of the largest and most comprehensive environmental epidemiologic studies of possible association between environmental factors and breast cancer. The Long Island study was ordered by Congress in 1993 in response to reports of elevated breast cancer deaths in a number of northeastern states. The researchers looked at DDT and its metabolite DDE, chlordane, dieldrin, and PCBs.

According to a report by the Environment News Service, although the study found no evidence that organochlorine compounds are associated with the elevated rates of breast cancer on Long Island, the researchers say that it is possible that breast cancer risk in some individuals may be associated with organochlorine exposures because of individual differences in metabolism and ability to repair DNA damage. "Recent research by other investigators suggests that organochlorine compounds may be related to the type of breast cancer that has clinical characteristics that are associated with decreased survival rates. This is an important issue that we are continuing to investigate among the women in our study," said principal investigator Dr. Marilie Gammon, associate professor of epidemiology at the University of North Carolina at Chapel Hill School of Public Health (Lazaroff 2002).

## POPS AND CARCINOGENESIS

Regarding the carcinogenic nature of dioxins, Joe Thornton has written,

Ironically, the most carcinogenic organochlorine is also the one whose carcinogenity is most controversial. TCDD (dioxin) is the most potent synthetic carcinogen ever tested in the laboratory. There have been 18 separate assessments of dioxin's carcinogenicity, involving five different species of experimental animals, both sexes, five routes of exposure, and high and low doses over short and long periods of time. In every case, dioxin has caused cancer. (Thornton 2000, 190)

The carcinogenic properties of dioxins have been accepted as valid science only since about 1990. Earlier than that, Monsanto, the primary manufacturer of many dioxins, emphatically denied that they caused cancer. As recently as the year 2000, the U.S. EPA was debating whether dioxin was carcinogenic, at the same time that the U.S. Army was giving a decoration (See chapter 1, "Agent Orange") to Vietnam veterans whose cancers could be traced to it.

Non-Hodgkin's lymphoma has been one of the most rapidly increasing malignant diseases in industrialized countries during the last two decades. In epidemiologic studies, non-Hodgkin's lymphoma has been associated with exposure to chemicals such as phenoxyacetic acids, chlorophenols, dioxins, organic solvents (including benzene), polychlorinated biphenyls, chlordanes, and immunosuppressive drugs. Levels of some types of dioxins and furans were found to be significantly higher in the adipose tissue of seven patients with malignant lymphoproliferative disease than in twelve surgical controls without malignant disease. In addition, the International Agency for Research on Cancer has established a strong link between cancers and exposure to dioxin. Experimental evidence and clinical observations indicate that these chemicals may impair the immune system, which may be related to the fact that risk for non-Hodgkins lymphoma increases in persons with acquired and congenital immune deficiency as well as autoimmune disorders ("Some Aspects" 1998).

## POPS AND CHILDREN: SPECIAL CONSIDERATIONS

Diazinon, one of the biggest-selling household pesticides in the United States, during December 2000 was ordered banned by the EPA in phases over a four-year period because of its health threats to children. Diazinon is widely sold under such brand names as Ortho, Spectracide, and Real-Kill. Diazinon is a member of the organophosphate family of chemicals, which attack the central nervous system and can be especially toxic to children, even in low doses. Retail sales for indoor use of the chemical have been ordered to cease by December 2002; sales for outdoor use are to end by December 2003. In December 2004 manufacturers will be obliged to repurchase unsold retail supplies.

A report by the Center for Health, Environment, and Justice revealed that nearly all Americans are exposed to unhealthy levels of dioxins through normal daily consumption of food. This exposure begins in the womb and continues through life. According to the report, "America's Choice: Children's Health or Corporate Profit?" children exposed to dioxins *in utero* during critical periods of development appear to be the most sensitive and vulnerable to their toxic effects. In children, dioxin exposure has been associated with lowered intelligence, increased prevalence of withdrawn and depressed behavior, adverse effects related to an ability to focus attention, an increase in hyperactive behavior, disrupted sexual development, birth defects, and damage to the immune system.

About 20 percent of an average mother's burden of fat-soluble chemicals may be transferred to a newborn child during six months of breast-feeding. Persistent organochlorines in a mother's body pass through the placenta to the embryo and fetus. Comment Ted Schettler and colleagues, "Breast-feeding infants may get a dose of dioxin, per kilogram of their body weight, within the first six months of life that exceeds health-based limits . . . five times the allowable daily intake of PCBs for a full-grown adult. Cow's milk with levels of PCBs this high would be too contaminated for sale in the United States" (Schettler et al. 1999, 205).

Exposure to dioxins occurs over a lifetime, and the danger is cumulative. Children's dioxin intake is proportionally much higher than adults' because of the chemical's presence in dairy products and breast milk. Levels of exposure that cause no noticeable effects in adults may cause irreversible harm to the developing fetus and newborn (Thomas and Colborn 1992). Pre- and postnatal exposure to persistent organochlorines may have profound effects in humans, among other animals. For example, lower birth weight, slower postnatal growth, and reduced short-term memory have been noted among the children of women who consumed moderate amounts of organochlorine-contaminated fish from Lake Michigan (Allsopp, Costner, and Johnston 1995; Jacobson and Jacobson 1990, 1993; Jacobson, Jacobson, and Humphrey 1990; Jacobson et al. 1992).

An average nursing infant receives doses of dioxins and PCBs that are twenty and fifty times higher than those of the average adult, respectively (Thornton 2000, 42). In a year of nursing, an average baby receives 4 to 12 percent of its total lifetime accumulation; because of bioaccumulation, this hypothetical infant will enter its second year of life with the dioxin or PCB levels of a seventy-year-old in the previous generation. These numbers are magnified again by the high levels of contaminants in certain geographical areas, such as the polar regions and higher elevations of mountain ranges. An Inuit's daily dose of dioxins, PCBs, toxaphene, and hexachlorobenzene, for example, range from twice to forty-eight times the Canadian government's "tolerable daily intake" for adults (Thornton 2000, 42). Following dioxin contamination at Times Beach, Missouri, during 1983, tests determined that children whose mothers lived near dioxin-contaminated areas during their pregnancies had abnormal brain function and altered immune systems. These effects, which were noted during early adolescence, are probably irreversible (Smoger et al. 1993).

A study has assessed the association between persistent organochlorine compounds acquired via consumption of fatty fish from the Baltic Sea (along the Swedish east coast) and infants' lower-than-normal birth weights. The study found that a mother's high current intake of fish from the Baltic Sea (at least four meals per month) tended to increase the risk of having an infant with low birth weight. Mothers who had grown up in a fishing village had an increased risk of having an infant with low birth weight (Rylander, Stromberg, and Hagmar 1996).

*In utero* exposure to polychlorinated biphenyls has been linked to adverse effects on neurological and intellectual function in infants and young children. Jacobson and Jacobson (1996) assessed these effects through school age and examined their importance in the acquisition of reading and mathematics skills. This study concluded that *in utero* exposure to PCBs in concentrations slightly higher than those in the general population can have a long-term impact on intellectual function. Developmental toxins may interfere with fetal development, sometimes causing birth defects. Neurotoxins can injure the developing brain, causing lifelong mental impairment.

During the late 1990s, government health ministries in Denmark and the Netherlands found that substantial amounts of phthalates are released from PVC into human saliva. Dioxins, highly potent synthetic hormone disrupters, are generated during the production and disposal of PVC. This is important because a number of children's teething toys have been manufactured from flexible PVC plastics. The governments of the Netherlands, Denmark, Belgium, Austria, and Spain have sought bans on the use of soft PVCs in toys. Some large toy retailers, including Toys-R-Us, removed these teething rings from their stores shortly thereafter (World Wildlife Fund 2000).

A reexamination of 212 children from the Lake Michigan Maternal Infant Cohort Study indicated neuro-developmental deficits assessed in infancy and early childhood persisted at age eleven (Jacobson and Jacobson 1996). These children were exposed *in utero* through mothers who consumed fish six months prior to and during pregnancy. After adjustment for a variety of potentially confounding factors, study results indicated the most highly exposed children (based on maternal milk PCB concentration) were three times as likely to have low full-scale and verbal IQ scores. They also were twice as likely to lag at least two years in reading comprehension. Many also had difficulty paying attention. "These intellectual impairments are attributed to *in utero* exposure to polychlorinated biphenyls and to related contaminants at concentrations slightly higher than those found in the general population" (U.S. EPA 1996).

The Lake Michigan study is striking not only because of the lasting impact seen in the children, but also because the fish-eating mothers were not highly contaminated. The levels measured in their bodies fall on the high end of what is considered the normal background range in the human population. The differences between the levels in children and their mothers reflect the power of biomagnification. The same study also is striking because it illustrates lasting impacts on the children.

In a similar U.S. study at the State University of New York (Oswego), researchers found measurable neurobehavioral deficits in newborn children of women who had eaten the equivalent of forty pounds of organochlorine-contaminated Lake Ontario salmon during a lifetime. These children showed abnormal reflexes, a shorter attention span, and an intolerance to stress. The Oswego study was the first to document a wide range of effects on temperament stemming from prenatal exposure to contaminants (World Wildlife Fund 2000).

A branch of the U.S. Public Health Service has concluded that PCBs and dioxins are responsible at least in part for neurological and behavioral deficits reported in children exposed in the womb. This assessment by the Agency for Toxic Substances and Disease Registry notes the "remarkable parallels" in the human epidemiological evidence and corroboration from wildlife and laboratory evidence: "The collective weight of the evidence indicates that certain PCB/dioxin-like compounds found in fish . . . can cause neurobehavioral deficits. Further, these compounds have produced some effects in some Great Lakes fish consumers" (World Wildlife Fund 2000).

A study in Mexico (Guillette et al. 1998, 347–53) reported striking differences in the development of children exposed to agricultural pesticides compared to children with minimal pesticide exposure. In this study, researchers tested two groups of four- and five-year-old children living in the Yaqui Valley region in northwestern Mexico. The two groups were similar in all respects, including ethnicity and diet, except for their exposure to pesticides. Families living in the foothills are ranchers who rely almost exclusively on traditional methods of pest control such as intercropping. The valley dwellers, on the other hand, live in an agricultural area that has seen heavy synthetic pesticide use since the 1940s.

Samples of human breast milk and cord blood taken from valley women contained high levels of POPs that included several targeted organochlorines: aldrin, endrin, dieldrin, heptachlor, and DDE. In tests developed to measure growth and development, the pesticide-exposed valley children fell far behind their foothill-dwelling peers. Over a period of a few years, the valley children exhibited decreased physical stamina in a jumping test, a lack of eye-hand coordination evident in their decreased ability to catch a ball, diminished memory, and a notable inability to draw a sketch of a person, which is used as a nonverbal measure of cognitive ability.

Louisiana and Texas emit more developmental and neurological toxins to air and water than any other states. Tennessee, Ohio, Illinois, Georgia, Virginia, Michigan, Pennsylvania, and Florida also are major emission sources (World Wildlife Fund 2000). "This is the first complete snapshot we've ever had of toxic pollution in this country that can affect the way that children's bodies and brains develop," said Jeff Wise, policy director of the National Environmental Trust (Gordon 2000). Roughly 12 million U.S. children under eighteen years of age (one out of every six) now suffer developmental, learning, or behavioral disabilities, including mental retardation, birth defects, autism, and attention-deficit hyperactivity disorder (Gordon 2000).

"Now we know what we have suspected for years, that toxic chemicals are bringing anguish to thousands of families in this country," said Larry B. Silver president of the Learning Disabilities Association of America and clinical professor of psychiatry at Georgetown University Medical Center. "These are families that worry, work overtime, and go without to take care of a child with a developmental or neurological disability like mental retardation or learning disabilities" (Gordon 2000).

This report (Gordon 2000) documents apparent increases in some birth defects, attention-deficit disorder, and autism. The report also compiles data describing increases in low birth-weight and premature births, both of which have been linked to effects of endocrine-disrupting chemicals in animal studies.

"While it's usually impossible to say that a particular child's disability is caused by a toxic chemical, it is clear that toxic chemicals are taking a tragic toll across the population," said Ted Schettler, an occupational and environmental health physician with Physicians for Social Responsibility. "This report is yet another in a series of wake-up calls to parents and policymakers that our children are being harmed by the current chemical environment and lack of regulatory oversight" (Gordon 2000).

## PCBS IN BREAST MILK

Research indicates that the breast milk of many women in all parts of the world now contains levels of PCBs high enough to damage a child's immune system. Such babies are several times more likely to contract chicken pox, middle-ear infections, and several other illnesses. The same research contradicts previous assurances by the British government, such as a comment by Sir Kenneth Calman, then (in 1997) the government's chief medical officer, who said PCBs posed "negligible threats" to human health (Dennis and Leake 2000).

A study led by Nynke Weisglas-Kuperus, a pediatrician at Sophia Children's Hospital in Rotterdam, Holland, examined the development of 207 babies born during the early 1990s. Half were breast-fed and half given formula. Weisglas-Kuperus found that many of the breast-fed babies received higher levels of PCBs than those fed formula. All other factors being equal, breast-fed babies usually suffer fewer illnesses than those who are bottle-fed, but high PCB levels negated this benefit.

In the United States and most other industrialized countries, PCBs are present in breast milk fat at about 1 ppm. An infant drinking such milk will take in a quantity of PCBs five times the allowable daily intake for an adult, according to standards established by the World Health Organization (Thomas and Colborn 1992, 365). Children exposed in the womb to PCBs at background levels in the United States have experienced hypotonia (loss of muscle tone) and hyporeflexia (weakened reflexes) at birth, delays in psychomotor development at ages six and twelve months, and diminished visual recognition memory at seven months (Tilson et al. 1990, 239).

Following cessation of commercial PCB production in the United States during the 1970s, PCB concentrations in breast milk initially declined. By the 1990s, however, PCB concentrations had generally stopped falling (Furst, Furst, and Wilmers 1994). Although production of many organochlorines in the United States has ceased, the ecosphere's burden of PCBs persists because quantities remain deposited in existing electrical equipment, in sediments, and in landfills and are still available for continued circulation throughout the global environment.

## PCBS AND SPONTANEOUS ABORTION

According to Rita Loch-Caruso of the Toxicology Program, Department of Environmental and Industrial Health, University of Michigan, some studies suggest that PCBs may disrupt the natural termination of pregnancy by stimulating premature uterine contractions. The cellular and molecular events underlying this toxicity are unknown. Loch-Caruso and other researchers at the University of Michigan have described a possible mechanism for PCB-induced uterine contractions. In a series of experiments they discovered that the PCB mixture aroclor 1242 can initiate cellular changes associated with contractions in uterine muscle cells.

Loch-Caruso contends that PCBs have been associated with spontaneous abortion and shortened gestation length in women, wildlife, and laboratory animals exposed to these industrial chemicals. Because premature birth is a considerable health problem in the United States (it accounts for 75 percent of newborn deaths not related to malformations), "it is important to determine how and to what extent PCBs may disrupt normal uterine functioning. The mechanistic understanding acquired in these studies is providing much needed insight into reproductive risks that may arise from PCB exposures" (Bae, Stuenkel, and Caruso 1999).

Increases in cancer mortality in workers exposed to PCBs have been observed in several workplace studies (Bertazzi, Riboldi, and Persatori 1987; Brown 1987; Sinks, Steele, and Smith1992; Yassi, Tate, and Fish 1994). Elevated risks of malignant melanoma, gastrointestinal tract cancer, liver cancer, gall bladder cancer, biliary tract cancer, and cancer of hematopoietic tissue have been reported following PCB exposure. Kuratsune and colleagues (1988) reported significant excess cancer of the liver by accidental ingestion of up to two grams of PCBs two decades before the tests.

Exposure to organic solvents in general, including the organochlorines tetrachloroethylene, trichloroethylene and 1,1,1-trichloroethane, has been found to be significantly associated with spontaneous abortion among female workers employed by industries using synthetic organochlorines. A study of Finnish pharmaceutical-industry workers employed between 1973 and 1981 revealed that exposure to organic solvents, particularly methylene chloride, was associated with an increased risk of spontaneous abortion. For example, exposure to four or more solvents was associated with a 350 percent increase in the risk of spontaneous abortion, which was statistically significant, and exposure to methylene chloride was associated with a 230 percent increase in the risk of spontaneous abortion, which had borderline significance. Furthermore the risk increased with increasing frequency of exposure to the solvents. The study also found an elevated risk for spontaneous abortion after exposure to estrogens when controlling for the effects of other chemicals, although the number of women studied was small (Taskinen, Lindbohm, and Hemminki 1986).

## DIETING DANGERS AND POPS

Researchers at Quebec City's Université Laval have found that weight loss may raise levels of PCBs and other pollutants in the bodies of dieters. Pierre Ayotte and Eric Dewailly, who have studied toxic organochlorine pollutants in the human body, joined with Jonathan Chevrier, Pascale Mauriege, Jean-Pierre Despres, and

Angelo Tremblay to test what happens to body burdens of these compounds when people lose weight. PCBs and other organochlorine compounds stored in fat tissue release their toxicity into the bloodstream as fat stores are dissolved. Such releases could cause "a rise of these compounds in the heart, lungs, kidneys, liver and spleen as well as brain" (Thompson 2001, A-3). These findings could help to explain why obese people who have lost weight sometimes have weakened immune systems and increased death rates.

"These results could possibly explain in part the controversial association between weight loss and increased mortality rates," the researchers wrote. "Further studies are needed in order to assess to what extent health complications and carcinomas could be triggered by weight loss through such an increase in plasma (blood) organochlorine concentrations" (Thompson 2001, A-3).

"It is far from clear that the increase we see after a weight loss brings about health problems," said Ayotte. He said that members of the research team are divided on the question. "I don't want people to get the idea that it's not good to lose weight" (Thompson 2001, A-3). Ayotte said that the concentrations of the chemicals in most Canadians are nowhere near the levels shown in populations such as the Inuit, or in countries such as Mexico where pesticides such as DDT are still in use. "To reach the levels found in the Inuit, a person would have to lose 80 percent of the fat in their body," Ayotte said (Thompson 2001, A-3). Ayotte said that the study indicates that Inuit who have ingested high levels of organochlorine compounds might have to be more careful in losing weight.

In the Laval study, the first of its kind on humans, the researchers followed thirty-nine obese individuals as they lost weight, testing their fat tissue and blood for twenty-six organochlorine compounds and compared them with a control group of fifty-seven average-weight women. The researchers found an increase in nineteen of the compounds studied, five of which increased significantly. Levels of toxics tended to rise as more weight was lost by any one individual.

## THE PHYSIOLOGICAL DANGERS OF DIOXINS

Dioxins are among the deadliest and most pervasive of synthetic chemical pollutants. Most dioxins are created by the application of heat to the melting point of various organochlorine products during several industrial processes, including the bleaching of most paper pulp and the production of many pesticides. Many household plastics (especially those containing PVCs) also release dioxins when burned, so small-scale (often rural) trash burning by individual households is a major source of these contaminants.

Exposure to dioxins and dioxin-like chemicals during fetal development, infancy, and early childhood may be eroding the physical and mental functions of current and future generations. Some of the health effects linked to dioxins and dioxin-like chemicals include neurodevelopmental effects, such as reduced IQ, increase in hyperactive behavior, adverse effects on attentional processes, increased prevalence of withdrawn or depressed behavior; altered immune function; central nervous system disorders; chloracne and other skin disorders; disrupted liver and kidney function; altered hormone levels, particularly thyroid, testosterone, and estrogen; reproductive effects such as altered sex ratio, reduced fertility; endometriosis; and liver, skin, and lung cancers ("Dioxin Deception" 2001).

A report by the EPA issued in 2000 concluded for the first time that dioxins are human carcinogens. For the first time, the agency's draft report classified the most potent form of dioxin—2,3,7,8-tetrachlorodibenzo-p-dioxin (TCDD)—as a human carcinogen, a step above the previous ranking of probable carcinogen. More than one hundred other dioxin-like compounds were classified as "likely" human carcinogens (Skrzycki and Warrick 2000, A-1).

Environmentalists, extrapolating from the EPA's risk findings, have estimated that about one hundred of the roughly fourteen hundred cancer deaths that occur daily in the United States are attributable to dioxins (Skrzycki and Warrick 2000, A-1). These findings came as a surprise even to EPA policymakers, who have tracked slowly falling levels of dioxins in the environment, following a series of tough new regulations on dioxin-emitting industries. The EPA said that industrial emissions of dioxins were reduced roughly 80 percent between 1987 and 1995.

The EPA report noted that emissions of dioxins have plummeted from their peak levels in the 1970s but still may pose a significant cancer threat to some people who ingest the chemical through foods in a normal diet. Standards for toxicity of dioxins (among other POPs) have been dropping in recent years to the point where some toxicologists believe that even residual amounts may have adverse health effects, especially (as is the case in much of the world today), because most people harbor some of these chemicals in their bodies. Dioxins have been linked to several cancers in humans, including lymphomas and lung cancer. The EPA report associates low-grade exposure to dioxins with a wide array of other health problems, including changes in hormone levels and developmental defects in babies and children. For a small segment of the population who eat large amounts of fatty foods, such as meats and dairy products that are relatively high in dioxins, the odds of developing cancer could be as high as one in 100, a risk ten times as high as the EPA's previous projections.

Dioxin is "the Darth Vader of toxic chemicals because it affects so many systems [of the body]," said Richard Clapp, a cancer epidemiologist at Boston University's School of Public Health (Skrzycki and Warrick 2000, A-1). During the last years of the 1990s, the U.S. EPA imposed regulations on major dioxin emitters, including municipal waste combustion, medical waste incinerators, hazardous waste incinerators, cement kilns that burn hazardous waste, and pulp and paper operations. When those regulations become fully effective over the next few years, the agency expects further declines of dioxin levels.

The production of dioxins permeates the manufacturing processes of many common daily products. For example, chlorine bleaching of pulp for paper products produces dioxins that can be found in pulps, waste sludge, and other effluent. Incomplete combustion during waste incineration, metals production, petroleum refinement, and fossil fuel combustion are major sources of dioxins. New methods in gas scrubbing and pulp bleaching have lowered the amount of chlorine by-products produced over the years (Zook and Rappe 1994).

## TOLUENE AND DEVELOPMENTAL TOXICITY

More than 98 million pounds of toluene were released into the air and water in the United States during 1998. The printing industry, with many small to medium-sized facilities near residential areas, is the largest source of air emissions of toluene, adding to the concern about children's exposure to this pollutant (Schettler 2000b). Toluene can cause abnormalities of the face and head resembling those of fetal alcohol syndrome, as well as growth retardation and persistent deficits in cognition, speech, and motor skills. These developmental toxins may interfere with normal fetal development, sometimes causing birth defects.

## THE DANGERS OF LINDANE

Lindane (g-HCH, hexachlorocyclohexane) is included in the government Red List of dangerous substances. It has been in use as a broad-range insecticide for fifty years, long enough to build up a significant body of evidence on its toxic and environmental hazards. The International Agency of Research on Cancer (IARC) has concluded that lindane is a possible human carcinogen (class 2B). The U.S. EPA has classified it similarly as a class B2/C possible human carcinogen. The presence of lindane in human milk has been reported in countries throughout the world.

Lindane has caused deaths and poisonings in humans and there is authoritative recognition of its long-term health effects including carcinogenesis. Scientific and anecdotal evidence links lindane with serious health problems, including aplastic anemia, birth disorders, and breast cancer (Pesticides Trust 1999). Included among the reported chronic effects of exposure to lindane are nervous disorders and increased liver weight.

Lindane is highly volatile. When it is applied to field crops, as much as 90 percent of the pesticide enters the atmosphere and is later deposited by rain. Lindane also leaches into surface and ground waters. Like other organochlorine pesticides, lindane is fat-soluble. This quality enhances its tendency to bioaccumulate along food chains. Residues have been detected in the kidneys, liver, and adipose tissue of a wide variety of wild animals and birds. It is highly toxic to aquatic invertebrates and fish.

Lindane had been banned or severely restricted in thirty-seven countries by 1999. The Advisory Committee on Pesticides in the United Kingdom has so far carried out three reviews of lindane and continued to recommend its approval. The Pesticides Trust believes that Lindane should be banned on the basis of existing evidence and as a precaution to avoid further health and environmental problems that are suspected of being caused by lindane (Pesticides Trust 1999).

Lindane is still widely used worldwide. It is commonly used as a treatment for lice and scabies in humans and also against ectoparasites in animals. In a control study, Davis and colleagues (1993) reported a statistically significant increase of brain cancer in children following treatment with lindane shampoo. Veterinary use in sheep can cause contamination of wool, as well as of milk and meat. Several cases of human poisoning by lindane have been reported. Children are significantly more susceptible than adults to the toxic effects of lindane. In one case a dose equivalent to 62.5 milligrams per kilogram proved fatal. In adults, doses above 300 milligrams per kilogram ingested orally have proved fatal.

Lindane also works as an endocrine disrupter that is capable of imitating hormones in humans and thereby disrupting the physiological functions that these hormones control. A significant body of evidence suggests that where lindane is used extensively, and particularly where cattle are exposed to it, the incidence of breast cancer rises. The United Kingdom has the highest rate of death from breast cancer in the world, and in Lincolnshire were lindane is used extensively on sugar-beet crops, the rate of breast cancer is 40 percent higher than the national average (Pesticides Trust 1999). Exposure to lindane also has been linked to blood dyscrasias as well as aplastic anemia, in which formation of platelets and white cells is disrupted (Pesticides Trust 1999).

## AXING ATRAZINE

Chemicals used to enhance agricultural yields also may cause problems with human health. During 1997, for example, the Environmental Working Group (Casey and Hayes 1999) said that levels of several herbicides, the most widely used of which is atrazine, were too high in the tap water of 245 midwestern cities and towns. In high doses, atrazine has been linked to several forms of cancer. In the report, titled "Weedkillers by the Glass," atrazine levels were said to be highest in cities and towns throughout the Corn Belt (Indiana westward to Nebraska) where atrazine is used liberally in surrounding agricultural areas. The report said that Omaha-area residents who drink water drawn from the Platte River increase their lifetime cancer risk by a factor of eleven. For those who drink water from the Missouri River (Omaha uses both rivers), the lifetime cancer risk is seven times the federal standard (Flanery 1994).

The makers of the chemicals protested. "The water is absolutely safe," said Chris Klose, speaking for the American Crop Protection

Association. "The [Environmental Working Group] study is without scientific merit, and it's damaging to the public trust" (Flanery 1994). The Environmental Working Group's study found atrazine in the drinking water of all the cities and towns tested in Nebraska and Iowa—Omaha, Des Moines, Davenport, Cedar Rapids, and Iowa City. The report recommended that parents in this area avoid exposing infants and young children to tap water between May 1 and August 30, when herbicide use on neighboring fields is most intense. Parents were urged to use bottled water for infant formula and frozen fruit juices. Donna Rhee, an environmental chemist in Omaha, said that risks from atrazine and other herbicides include breast cancer, gland cancer, and chromosome damage in animals at levels that human beings have been ingesting (Thomas 1995).

## UNINTENDED CONSEQUENCES OF WATER CHLORINATION

Chlorine (hypochlorite) is added to drinking water in many countries as a disinfectant to prevent infectious diseases. It has proved to be an enormously successful health-care measure, doubtlessly preventing widespread illness and death. Chlorination of water may provoke formation of several volatile organic chlorinated compounds as the chlorine reacts with organic matter in the water. Most of these by-products are called trihalomethanes (THM) and include chloroform, which is known to cause cancer in animals (Allsopp, Costner, and Johnston 1995; Cantor 1993; Zieler et al. 1988).

Evidence from several epidemiology studies has suggested an association between chlorination of drinking water and an increased risk of cancers of the bladder, colon, and rectum (Cantor 1993). Which chlorinated by-products in water are responsible for the increased risk of cancers is not yet clear. Recent animal studies have suggested that the organic by-products of chlorination, namely THMs, are of greatest concern (Dunnick and Melnick 1993). Evidence also indicates, however, that the nonvolatile chlorinated by-products are carcinogenic and are responsible for a major part of the toxicity (Cantor 1993).

For individuals using chlorinated swimming pools, inhalation is the main source of exposure, and since the chlorinated by-products are volatile, particularly high exposures are experienced by those who swim for a long time while exercising vigorously, which accelerates metabolism. Studies on competition swimmers using indoor chlorinated swimming pools have found that chloroform levels were elevated in their blood (Aiking et al. 1994).

## CHLORPYRIFOS

Chlorpyrifos (better known under its trade name, Dursban) is a key ingredient in a variety of products such as pet flea collars, ant sprays, and, most notably, products designed for termite control. Chlorpyrifos belongs to a class of thirty-seven persistent pesticides known as organophosphates, which were initially developed as nerve gases during World War II by the German chemical giant I.G. Farben. Recent studies at Rutgers University indicate that chlorpyrifos persists much longer indoors than had been previously recognized. Carpets, soft furniture, and plush toys are especially likely to absorb chlorpyrifos and to retain it for long periods of time as vapors are released into the air.

Chlorpyrifos has been used widely for more than thirty years in agriculture and in hundreds of products utilized by exterminators and homeowners, including some Raid sprays, Hartz Yard and Kennel Flea spray, and Black Flag Liquid Roach and Ant Killer. An EPA assessment found that the chemical could damage the brain in newborn rats and cause weakness, vomiting, diarrhea, and other ill effects in children.

Increased concern about chlorpyrifos emerged after studies—some conducted by its manufacturer—were released showing that the compound causes brain damage in fetal rats whose mothers consumed the pesticide. The results of the animal tests are considered serious enough to indicate that the pesticide should not be used where there are children, although the direct link to humans is yet to be established.

The U.S. EPA has announced that products containing chlorpyrifos will be phased out for home and garden use. The new rules allow continued use of chlorpyrifos on many crops but sharply limit its use on apples, grapes, and tomatoes. The new regulations also entirely eliminate its use around homes, schools, day-care centers, and other places where children may be exposed.

Restrictions on chlorpyrifos developed as part of the EPA's reexamination of organophosphate pesticides. In 1999 the EPA banned the use of pesticides with organophosphates and methyl parathion on fruits and many vegetables and restricted the use of azinphos-methyl. The Food Quality Protection Act of 1996 called for a much stricter appraisal of chemical risks, focusing on potential harm to children, whose small bodies and fast-growing brains are particularly vulnerable to toxic chemicals (Revkin 2000).

Indian environmentalists during 2002 initiated a campaign for an immediate ban on use of endosulfan, a pesticide that previously had been outlawed in many parts of the world. The campaign followed an Indian government report that linked use of endosulfan with disease and deformity, particularly of infants, in southern India. According to a report by the Environment News Service, activists of the New Delhi–based group Centre for Science and Environment (CSE) called for reinstatement of a former ban in the southern Indian state of Kerala ("Indian Enviros" 2002).

The CSE claimed to have obtained a copy of the report from unnamed sources in Kerala. The report found endosulfan residues in water and blood samples collected from Padre, a village in northern Kerala, in September and October 2001, ten months after the pesticide had last been sprayed on the region's cashew crops. The report was said to have found a "significantly higher prevalence of learning disabilities, low I.Q. and scholastic backwardness" among children, as well as problems and congenital and reproductive abnormalities among people in the region. "It has also been found that workers in the cashew plantations suffer from neurological problems such as trembling hands," said Kushal Pandey, a reporter who has been following the issue up for CSE's environment magazine, *Down to Earth* ("Indian Enviros" 2002).

## MTBE: COMING TO A WATER TAP NEAR YOU

Human-created chemicals sometimes are introduced as solutions to environmental problems, only to end up being problems themselves. Such has been the case with a common gasoline additive, methyl tertiary butyl ether (MTBE), which has been used since about 1990 to reduce the amount of pollution delivered to the atmosphere by automobiles burning gasoline.

The compound was never adequately tested before it was marketed. When MTBE was enlisted in the war on automotive air pollution, its effects on human health were unknown. The compound helped clean the air, true enough. Only a few years after it began leaking from thousands of faulty gasoline storage tanks around the United States did anyone realize what it was doing to the water. By the early 1990s, MTBE had become an important groundwater contaminant.

MTBE, derived from natural gas, is added to gasoline in many parts of the United States to reduce carbon monoxide and ozone levels in the air (95 percent of total MTBE use), or to increase the octane of gasoline (5 percent of use). MTBE is the most widely used oxygenate for these purposes; however, ethanol is used in many areas of the United States. Oxygenates such as MTBE and ethanol reduce the need for benzene and other ozone-forming aromatic compounds in gasoline.

MTBE, which was developed during the 1970s to replace lead as an octane-booster for gasoline, is called an oxygenate because it adds oxygen to gasoline. MTBE smells like turpentine and spreads so quickly and evenly in an aquatic environment that a tablespoon is more than enough to spoil the amount of water in an Olympic-sized swimming pool. Even at five parts per billion, MTBE has a distinct smell. MTBE is suspected of causing cancer in animals. (A European study during the middle 1990s linked MTBE to liver and kidney tumors in mice.)

In the South Tahoe Public Utility District, twelve of thirty-four wells had been forced to close by early 2000 because of MTBE contamination. "It's a diabolical chemical. It moves up. It moves down. It moves everywhere. Our feeling is that as long as MTBE is in gasoline, our groundwater is in jeopardy," said Dennis Cocking of the South Lake Tahoe Public Utility District ("Additive Poses" 2000).

During April 2002 a San Francisco jury determined that gasoline containing MTBE is a defective product and that two major oil companies knew about the problem when they began marketing fuels containing the additive. The verdict, issued April 15, came in a product-liability case filed by the South Lake Tahoe Public Utility District over contamination of the district's groundwater. According to a report by the Environment News Service, MTBE had forced the District to close one-third of its drinking-water wells by 1998, when the utility filed suit ("Jury Labels" 2002).

According to the ENS account, "Lyondell Chemical Company, formerly Atlantic Richfield Chemical Company, Shell Oil Company, and Tosco Corporation, now part of Phillips Petroleum, knew their product was defective but withheld that information from the public

when they started selling gasoline boosted with methyl tertiary butyl ether (MTBE)" ("Jury Labels" 2002). This verdict is the first of its
kind, but dozens of similar cases have been filed by other utilities,
communities, and individuals across the country. In the next phase
of the trial, the jury will determine whether MTBE from the three oil
companies caused the groundwater pollution in South Lake Tahoe;
if the companies are found directly liable, damages will be assessed.

Santa Monica, California, also detected MTBE in its water supply.
By 2000 California (consumer of a tenth of the MTBE used in the
United States) had banned the substance within its borders after
2003. In the meantime, manufacturers of MTBE defended its use
as an environmental asset: "Because of cleaner-burning gasoline
with MTBE, cities like Los Angeles are enjoying their best air quality
in 50 years," said Terry Wigglesworth, executive director of the Ox-
ygenated Fuels Association ("Additive Poses" 2000).

Paul Squillace, a U.S. Geological Survey research hydrologist,
said the detection of MTBE varied substantially among the thirteen
urban areas investigated. Urban areas where MTBE was most commonly detected included Denver (79 percent of wells), Harrisburg,
Pennsylvania (37 percent), and various cities in New England (37
percent). Urban areas where MTBE was not detected include Ocala
and Tampa, Florida, Portland, Oregon, and Virginia Beach, Virginia.
Possible sources of MTBE in groundwater include leaking storage
tanks and nonpoint sources such as recharge of precipitation and
stormwater runoff (Squillace, Pope, and Price 1995; U.S. Geological
Survey 1997). By contrast, only 1.3 percent of the wells sampled in
twenty agricultural areas had detectable concentrations of MTBE.
The U.S. Geological Survey said that none of the urban wells sampled was being used as a source of drinking water. In general, public
water supplies draw from deeper groundwater, where MTBE contamination is less likely.

The amount of MTBE released during refueling at service stations
and from engine exhaust is unknown, but it probably constitutes
an important source of MTBE contamination. Leaking underground
storage tanks and spills also may be sources of MTBE pollution.
Although MTBE will vaporize from soils, it can move into groundwater. Once in groundwater, MTBE is more resistant to decay than
other gasoline components such as benzene (Snow and Zogorski
1995).

Ethanol (a gasoline-and-corn mixture widely used in the American Midwest) performs many of the same functions as MTBE but

costs more to produce. By the year 2000, some oil-industry leaders were making the switch despite the cost, for environmental reasons. Tosco, which owns Union 76 and Circle K stations, in December 2001 began to retool its refineries to utilize ethanol instead of MTBE. "Ethanol is a renewable fuel with great environmental benefits, including significant reductions in emissions of global warming gasses," said environmentalist Elisa Lynch of the Bluewater Network (Gardner 2000). "Farmers and the entire state would benefit from converting biomass to renewable fuel while we're addressing the concerns of how to dispose of agricultural wastes," said Jack King, national affairs manager for the California Farm Bureau (Gardner 2000).

Shell Oil Company during August 2002, agreed to pay $28 million to settle a lawsuit over MTBE contamination of wells in the South Tahoe Public Utility District. Pollution with MTBE had forced the South Lake Tahoe District to close one-third of its drinking-water wells by 1998, when the utility filed suit. The South Lake Tahoe Public Utility District also has settled with Chevron, Exxon, and others, and is expected, with the Shell settlement, to receive more than $69 million from refiners for cleanup and legal costs ("Shell Will Pay" 2002).

In April 2002, a California jury decided that gasoline containing MTBE is a "defective product," and that Shell, Lyondell Chemical Company, and Tosco Corporation knew about the problem and withheld information from the public when they began marketing fuels containing MTBE. During July 2002, MTBE manufacturer Lyondell agreed to pay $4 million to settle its role in the South Lake Tahoe case. Before the case went to trial, Exxon settled for $12 million and Chevron settled for $10 million ("Shell Will Pay" 2002).

The settlements could set a precedent for similar cases now being tried in other states across the nation. At least sixteen states have reported water contamination due to MTBE. According to the Environment News Service, California Governor Gray Davis has issued a ban on the use of MTBE in all fuel sold within the state, effective January 2004. At least three oil companies (Atlantic Richfield, Exxon, and Shell) have said that they plan to stop boosting their gasoline with MTBE before that deadline ("Shell Will Pay" 2002).

### REFERENCES

"Additive Poses Hard Choice: Clean Air or Clean Water?" 2000. *Omaha World-Herald,* January 26.

Aiking, H., M. B. van Acker, R. J. P. M. Scholten, J. F. Feenstra, and H. A. Valkenburg. 1994. "Swimming-Pool Chlorination: A Health Hazard?" *Toxicological Letters* 72: 375–80.

Allsopp, Michelle, Pat Costner, and Paul Johnston. 1995. *Body of Evidence: The Effects of Chlorine on Human Health.* London: Greenpeace International, 1995. http://www.greenpeace.org/~toxics/reports/recipe.html.

Anderson, Julie. 2000. "Diazinon Sales to Be Eased Out." *Omaha World-Herald,* December 6.

Bae, J., E. L. Stuenkel, and R. Loch-Caruso. 1999. "Stimulation of Oscillatory Uterine Contraction by the PCB Mixture Aroclor 1242 May Involve Increased [Ca2 + ](i) through Voltage-Operated Calcium Channels." *Toxicology and Applied Pharmacology* 155, no. 3: 261–72.

Bertazzi, P. A., A. L. Riboldi, A. and A. Persatori. 1987. "Cancer Mortality of Capacitor Manufacturing Workers." *American Journal of Industrial Medicine* 11: 65–176.

Brown, D. P. 1987. "Mortality of Workers Exposed to Polychlorinated Biphenyls: An Update." *Archives of Environmental Health* 42, no. 6: 333–39.

Cantor, D. S., G. Holder, W. Cantor, P. C. Kahn, G. C. Rodgers, G. H. Smoger, W. Swain, H. Berger, and S. Suffin. 1993. "*In-utero* and Postnatal Exposure to 2,3,7,8-TCDD in Times Beach, Missouri: Impact on Neurophysiological Functioning." Paper presented at Dioxin '93, Thirteenth International Symposium on Chlorinated Dioxins and Related Compounds, Vienna, September 20–24.

Carson, Rachel. *Silent Spring.* Boston: Houghton-Mifflin, 1962.

Casey, Mike, and Melissa Hayes. 1999. "Into the Mouths of Babes: Government Underestimates Infant Exposure to Toxic Weed Killer." Press release, Common Dreams, Washington, D.C. July 28. http://www.commondreams.org/pressreleases/july99/072899d.htm.

Chen, Y.-C. J., Y.-L. Guo, and C.-C. Hsu. 1992. "Cognitive-Development of Yu-cheng (Oil Disease) Children Prenatally Exposed to Heat-Degraded PCBS." *Journal of the American Medical Association* 268: 3213–18.

Colborn, T., D. Dumanoski, and J. P. Myers. 1996. *Our Stolen Future: Are We Threatening Our Fertility, Intelligence, and Survival? A Scientific Detective Story.* New York: Penguin.

Cook, Judith, and Chris Kaufman. 1982. *Portrait of a Poison: The 2,4,5-T Story.* London: Pluto Press.

Davis, D. L., H. L. Bradlow, M. Wolff, T. Woodruff, D. G. Hoel, and H. Anton-Culver. 1993. "Medical Hypothesis: Xenooestrogens As Preventable Causes of Breast Cancer." *Environmental Health Perspectives* 101, no. 5: 372–77.

Dellinger, J. A., N. Kmiecek, and S. Gerstenberger. 1995. "Mercury Contamination of Fish in the Ojibwa Diet: 1. Walleye Fillets and Skin-on versus Skin-off Sampling." *Water, Air, and Soil Pollution* 80: 69–76.

Dennis, Guy, and Jonathan Leake. 2000. "Breast-feeding Mothers May Pass Toxins to Babies." *London Times,* April 30.

"Dioxin Deception: How the Vinyl Industry Concealed Evidence of Its Dioxin Pollution." 2001. Greenpeace USA. March 27. http://www.green peaceusa.org/toxics/dioxin_deceptiontext.htm.

Dunnick, J. K., and R. L. Melnick. 1993. "Assessment of the Carcinogenic Potential of Chlorinated Water: Experimental Studies of Chlorine, Chloramine, and Trihalomethanes." *Journal of the National Cancer Institute* 85, no. 10: 817–23.

Fein, G. G., J. L. Jacobson, and S. W. Jacobson. 1984. "Prenatal Exposure to Polychlorinated Biphenyls: Effects on Birth Size and Gestation Age." *Journal of Pediatrics* 105: 315–20.

Fitzgerald, E. F., K. A. Brix, D. A. Deres, S. A. Hwang, B. Bush, G. L. Lambert, and A. Tarbell. 1996. "Polychlorinated Biphenyl (PCB) and Dichlorodiphenyl Dichloroethylene (DDE) Exposures among Native American Men from Contaminated Great Lakes Fish and Wildlife." *Toxicology and Industrial Health* 12: 361–68.

Fitzgerald, E. F., S. Hwang, K. A. Brix, B. Bush, J. Quinn, and K. Cook. 1995. "Exposure to PCBs from Hazardous Waste among Mohawk Women and Infants at Akwesasne." Report for the Agency for Toxic Substances and Disease Registry, Atlanta.

Flanery, James Allen. 1994. "Debate on Water Re-ignites: Herbicides and Risk of Cancer Reported." *Omaha World-Herald,* October 19.

Furst, P., C. Furst, and K. Wilmers. 1994. "Human Milk As a Bio-indicator for Body Burden of PCDDs, PCDFs, Organochlorine Pesticides, and PCBs." *Environmental Health Perspectives: Supplements* 102, suppl. 1: 187–93.

Gardner, Michael. 2000. "Gas Refiner Replaces Disputed Additive; MTBE Is Out, Ethanol Soon to Be in for Tosco." *San Diego Union,* December 22.

Gordon, Anita. 2000. "New Report Concludes Nation Is Awash in Chemicals That Can Affect Child Development and Learning: Louisiana, Texas Emissions Lead the Country in First Effort Ever to Assess Scope and Sources of Developmental and Neurological Toxin Pollution; Report Documents Disturbing Trends in Developmental and Learning Deficits." Press release, Physicians for Social Responsibility, Washington, D.C. September 7. http://www.psr.org/trireport.html.

Guillette, Elizabeth A., Maria Mercedes Meza, Maria Guadalupe Aquilar, Alma Delia Soto, and Idalia Enedina Garcia. 1998. "An Anthropological Approach to the Evaluation of Preschool Children Exposed to Pesticides in Mexico." *Environmental Health Perspectives* 106, no. 6 (June): 347–53. http://www.anarac.com/elizabeth_guillette.htm.

Harada, M. 1976. "Intrauterine Poisoning: Clinical and Epidemiological Studies and Significance of the Problem." *Bulletin of the Institute of Constitutional Medicine* (Kumamato University, Japan) 25, suppl.: 169–84.

Hsu, S. T., C. I. Ma, S. K. Hsu, S. S. Wu, N. H. M. Hsu, C. C. Yeh, and S. B. Wu. 1985. "Discovery and Epidemiology of PCB Poisoning in Taiwan: A Four-Year Follow-up." *Environmental Health Perspectives* 59: 5–10.

Humphrey, H. E. B. 1983. "Population Studies of PCBs in Michigan Residents." In *PCBs: Human and Environmental Hazards*, edited by F. M. D'Itri and M. Kamrin. Boston, Mass.: Butterworth.

"Indian Enviros Urge Ban on Pesticide Endosulfan." 2002. Environmental News Service. July 3. http://ens-news.com/ens/jul2002/2002-07-03-02.asp.

Jacobson, J. L., G. G. Fein, S. W. Jacobson, P. M. Schwartz, and J. K. Dowler. 1985. "The Effect of Intrauterine PCB Exposure on Visual Recognition Memory." *Child Development* 56: 856–60.

Jacobson, J. L., and S. W. Jacobson. 1990. "Effects of Exposure to PCBs and Related Compounds on Growth and Activity in Children." *Neurotoxicology and Teratology* 12: 319–26.

———. 1993. "A Four-Year Follow-up Study of Children Born to Consumers of Lake Michigan Fish." *Journal of Great Lakes Research* 19, no. 4: 776–83.

———. 1996. "Intellectual Impairment in Children Exposed to Polychlorinated Biphenyls *in Utero*." *New England Journal of Medicine* 335, no. 11: 783–89.

Jacobson, J. L., S. W. Jacobson, and H. E. B. Humphrey. 1990. "Effects of *in Utero* Exposure to Polychlorinated Biphenyls and Related Contaminants on Cognitive Functioning in Young Children." *Journal of Pediatrics* 116: 38–45.

Jacobson J. L., S. W. Jacobson. R. J. Padgett, G. A. Brumitt, and R. L. Billings. 1992. "Effects of Prenatal PCB Exposure on Cognitive Processing Efficiency and Sustained Attention." *Developmental Psychology* 28: 297–306.

"Jury Labels MTBE Gasoline As Defective Product." 2002. Environment News Service, April 18. http://ens-news.com/ens/apr2002/2002L-04-18-09.html#anchor3.

Kreiss, K., M. M. Zack, and R. D. Kimbrough. 1981. "Association of Blood Pressure and Polychlorinated Biphenyl Levels." *Journal of the American Medical Association* 245: 2505–9.

Kuratsune, M., M. Ikeda, Y. Nakamura, and T. Hirohata. 1988. "A Cohort Study on Mortality of Yusho Patients: A Preliminary Report." In *Un-*

*usual Occurrences As Clues to Cancer Etiology*, edited by R. W. Miller. Tokyo: Japan Scientific Society Press/Taylor and Francis.

Lazaroff, Cat. 2002. "No Link Between Organochlorines, Breast Cancer." *Environment News Service*. August 6. http://ens-news.com/ens/aug 2002/2002-08-06-06.asp.

Lonky, E. J., T. Reihman, T. Darvill, J. Mather, and H. Daly. 1996. "Neonatal Behavioral Assessment Scale Performance in Humans Influenced by Maternal Consumption of Environmentally Contaminated Lake Ontario Fish." *Journal of Great Lakes Research* 22, no. 2: 198–212.

Montague, Peter. 1994. "PCBs Diminish Penis Size." *Rachel's Environment and Health News* 372, January 13. http://www.rachel.org/bulletin/index.cfm?St = 2.

Moyers, Bill. 2001. "Trade Secrets: A Moyers Report." Program transcript, Public Broadcasting Service, March 26. http://www.pbs.org/trade secrets/transcript.html.

Napier, Robert. 2001. "Hot Air on the Environment." *Guardian* (London), August 16.

Pesticides Trust. 1999. "Persistent Organic Pollutants and Reproductive Health." From a briefing for UNISON prepared by the Pesticides Trust, London. http://irptc.unep.ch/pops/default.html.

Revkin, Andrew C. 2000. "EPA Sharply Curtails the Use of a Common Insecticide." *New York Times*, June 9. http://irptc.unep.ch/pops/new layout/press_items.htm.

Ries, Lynn A. G., Phyllis A. Wingo, Daniel S. Miller, Holly L. Howe, Hannah K. Weir, Harry M. Rosenberg, Sally W. Vernon, Kathleen Cronin, and Brenda K. Edwards. 2000. "The Annual Report to the Nation on the Status of Cancer, 1973–1997, with a Special Section on Colorectal Cancer." *Cancer* 88, no. 10: 2398–2424. http://seer.cancer.gov/publications/csr1973_1998/overview.

Rupa, D. S., P. P. Reddy, and O. S. Reddy. 1991. "Reproductive Performance in Population Exposed to Pesticides in Cotton Fields in India." *Environmental Research* 55: 123–28.

Rylander, Lars, Ulf Stromberg, and Lars Hagmar. 1996. "Dietary Intake of Fish Contaminated with Persistent Organochlorine Compounds in Relation to Birth-Weight." *Scandinavian Journal of Work and Environmental Health* 22: 260–66.

Safe, S. H. 1994. "Polychlorinated Biphenyls (PCBs): Environmental Impact, Biochemical and Toxic Responses, and Implications for Risk Assessment." *Critical Reviews in Toxicology* 24, no. 2: 87–149.

———. 1995. "Environmental and Dietary Oestrogens and Human Health: Is There a Problem?" *Environmental Health Perspectives* 103, no. 4: 346–51.

Schantz, S. L., A. M. Sweeney, J. C. Gardiner, H. E. B. Humphrey, R. J. McCaffrey, D. M. Gasior, K. R. Srikanth, and M. L. Budd. 1996. "Neuropsychological Assessment of an Aging Population of Great Lakes Fisheaters." *Toxicology and Industrial Health* 12: 403–17.

Schettler, Ted. 2000a. "The Precautionary Principle and Persistent Organic Pollutants." The Science and Environmental Health Network. March. http://www.alphacdc.com/ien/pops_precautionary_ted.html.

———. 2000b. "Statement of Ted Schettler, MD, Physicians for Social Responsibility." Press conference, Willard Hotel, Washington, D.C., September 7. http://www.psr.org/trited.html.

Schettler, Ted, Gina Solomon, Maria Valenti, and Anne Huddle. 1999. *Generations at Risk: Reproductive Health and the Environment.* Cambridge, Mass.: MIT Press.

"Shell Will Pay $28 Million to Clean Wells of MTBE." Environment News Service. August 6. http://ens-news.com/ens/aug2002/2002-08-06-09.asp.

Sinks, T., G. Steele, and A. B. Smith. 1992. "Mortality among Workers Exposed to Polychlorinated Biphenyls." *American Journal of Epidemiology* 136, no. 4: 389–98.

Skrzycki, Cindy, and Joby Warrick. 2000. "EPA Links Dioxin to Cancer: Risk Estimate Raised Tenfold." *Washington Post,* May 17. http://irptc.unep.ch/pops/newlayout/press_items.htm.

Smoger, G. H., P. C. Kahn, G. C. Rodgers, and S. Suffin. 1993. "*In-utero* and Postnatal Exposure to 2,3,7,8-TCDD in Times Beach, Missouri." *Organohalogen Compounds* 13: 345–48.

Snow, Mitch, and John Zogorski. 1995. "Gasoline Additive Found in Urban Ground Water." United States Geological Survey, March 31. http://sd.water.usgs.gov/nawqa/vocns/mtbe.htm.

"Some Aspects of the Etiology of Non-Hodgkins Lymphoma." http://ehp net1.niehs.nih.gov/docs/1998/Suppl-2/679–681hardell/abstract.html.

Squillace, Paul J., Daryll A. Pope, and Curtis V. Price. 1995. "Occurrence of Gasoline Additive MTBE in Shallow Ground Water in Urban and Agricultural Areas" U.S. Geological Survey Fact Sheet 114–95. http://wwwrvares.er.usgs.gov/nawqa/nawqa_home.html.

Taskinen, H., M-L. Lindbohm, and K. Hemminki. 1986. "Spontaneous Abortion among Women Working in the Pharmaceutical Industry." *British Journal of Industrial Medicine* 43: 199–205.

Thomas, Fred. 1995. "Clear-cut Answers on Safety of Omaha's Drinking Water in Short Supply." *Omaha World-Herald,* August 27.

Thomas, K. B., and T. Colborn. 1992. "Organochlorine Endocrine Disrupters in Human Tissue." In *Chemically-Induced Alterations in Sexual and Functional Development: The Wildlife/Human Connection,* edited by T. Colborn and C. Clement. Princeton, N.J.: Princeton Scientific Publishing.

Thomas, P. T. 1990. "Approaches Used to Assess Chemically Induced Impairment of Host Resistance and Immune Function." *Toxic Substances Journal* 10: 241–78.

Thompson, Elizabeth. 2001. "A Slimmer You May Be Less Healthy: Quebec Researchers Find Link between Weight Loss and Higher Levels of Pollutants in the Body." *Montreal Gazette,* January 29.

Thornton, Joe. 2000. *Pandora's Poison: Chlorine, Health, and a New Environmental Strategy.* Cambridge, Mass.: MIT Press.

Tilson, Hugh A., et al. 1990. "Polychlorinated Biphenyls and the Developing Nervous System: Cross-Species Comparisons." *Neurotoxicology and Teratology* 12: 239–48.

U.S. Environmental Protection Agency. 1996. *Public Health Implications of PCB Exposures.* Atlanta: Agency for Toxic Substances and Disease Registry.   http://www.epa.gov/region5/foxriver/lower_fox_river_PCB_ Exposures.htm.

U.S. Geological Survey. 1997. "MTBE in Ground Water." April 15. http:// sd.water.usgs.gov/nawqa/vocns/mtbe.htm.

Waller, D. P., C. Presperin, M. L. Drum, A. Negrusz, A. K. Larsen, H. van der Ven, and J. Hibbard. 1996. "Great Lakes Fish As a Source of Maternal and Fetal Exposure to Chlorinated Hydrocarbons." *Toxicology and Industrial Health* 12: 335–45.

Wong, K. C., and M. Y. Huang. 1981. "Children Born to PCB-Poisoned Mothers." *Clinical Medicine* 7: 83–87.

World Wildlife Fund. 2000. "Toxics—What's New." http://www.worldwild life.org/toxics/whatsnew/pr_7.htm..

Yao, Y., A. Hoffer, C. Chang, and A. Puga. 1995. "Dioxin Activates HIV-1 Gene Expression by an Oxidative Stress Pathway Requiring a Functional Cytochrome P450 CYP1A1 Enzyme." *Environmental Health Perspectives* 103, no. 4: 366–71.

Yassi, A., R. Tate, and D. Fish. 1994. "Cancer Mortality in Workers Employed at a Transformer Manufacturing Plant." *American Journal of Industrial Medicine* 25, no. 3: 425–37.

Zieler, S., L. Feingold, R. A. Danley, and G. Craun G. 1988. "Bladder Cancer in Massachusetts Related to Chlorinated and Contaminated Drinking Water: A Case-Control Study." *Archives of Environmental Health* 43, no. 2: 195–200.

Zook, D. R., and C. Rappe. 1994. "Environmental Sources, Distribution and Fate of Polychlorinated Dibenzodioxins, Dibenzofurans, and Related Organochlorides." In *Dioxins and Health,* edited by A. Schecter. New York: Plenum.

# 7

# Toxic Barbie? Not Your Great-Grandmother's Estrogen

The density and abundance of sperm required to maintain the fecundity of the human animal is open to dispute; at some point, however, damage to sperm counts, taken far enough, must inhibit reproduction. Comes now the most intriguing question of our journey into the endocrine-disrupting potential of POPs: Has humankind's ingenuity—notably our products' dexterity in mimicking estrogen—planted the seeds of an eventual reduction in humankind's population? And what could such an outcome do to many a cherished assumption of the early twenty-first century, with increasing numbers of human beings crowded onto an Earth beset by an overload of greenhouse gases and other forms of human-induced pollution?

Eventually, perhaps, a cold wind may blow through the halls of human artifice. Amid humanity's festival of fecundity now come warnings that synthetic chemicals may prejudice the reproductive systems of many animals, including human beings. One report commented, "This is of great concern because effects on the next generation, and especially disturbances of the reproductive system, rapidly threaten populations as a whole" (Allsopp, Costner, and Johnston 1995).

We come now to the era of competing anthropogenic ecological holocausts. For example, could the endocrine disrupters in many man-made chemicals cause human sperm counts to fall so low that the human race will eventually find its gene pool in peril? Such a future could certainly (if eventually) stabilize the atmosphere's overload of greenhouse gases, for example. The endocrine-disrupting qualities of many organochlorines pose one of the fundamental ironies of global pollution. Has the ingenuity of the chemical industry created products that will ultimately render sterile at least some portions of the human race? Eventually, could the human race, author of so many plant and animal extinctions because of its prolific breeding and expropriation of the Earth's resources, itself become a victim of human artifice?

H. Burlington and V. F. Lindeman were the first, in 1950, to suggest that DDT might mimic estrogen, "because of its inhibition of male characters in developing cockerels" (Jefferies 1975, 193, 224). Burlington and Lindeman used seventy male chicks (thirty controls and forty experimental animals) to investigate the developmental effects of DDT. The experimental animals were injected with purified DDT in a solvent of chicken fat; the control animals were injected with chicken fat alone. According to Sheldon Krimsky, "The results were striking. The experimental animals had smaller testes and showed arrested development of secondary sex characteristics, compared with the controls" (Krimsky 1999, 6).

Colborn and colleagues have argued that

Evidence already exists that a number of organochlorine chemicals (such as dioxin, PCBs, and DDT) have reached concentrations in aquatic food sources that can lead to substantial functional deficits in animals that consume this food. . . . Based on current breast-milk concentrations nationwide, it is estimated that at least 5 percent and possibly more of the babies born in the United States are exposed to quantities of PCBs sufficient to cause neurological effects. (Colborn, vom Saal, and Soto 1993, 378)

As of this writing, however, any prospective, notable decline in human numbers remains a matter of speculation. During the year 2000 on the Christian calendar, half the human beings who ever had lived were alive, six billion people. Extinction of the human race does not seem to be a pressing issue, given our inundation of the world with increasing tides of humanity. At the same time, some estimates of male sperm counts have declined by half in the United States and other parts of the world during the half century since DDT and other synthetic organochlorines were first introduced into the environment (Sharpe 1993).

We may be skewing the natural order of reproduction by bathing the biosphere in synthetic estrogens. By the middle 1990s, at least 51 synthetic chemicals had been detected that disrupt the endocrine system in some way (Colborn, Dumanoski, and Myers 1996, 81). Some, like DES, mimic natural estrogen, while others interfere with testosterone, thyroid metabolism, or other parts of the system. This list of endocrine disrupters includes (as one of the 51 previously mentioned) the 209 different compounds that are classed as PCBs, 75 dioxins, and 135 furans (Colborn, Dumanoski, and Myers 1996, 81).

During April 1997 the European Environment Agency published a report on a major conference on endocrine-disrupting chemicals held December 2–4, 1996, in Weybridge, England. The Weybridge Report concluded

It is evident that there are adverse health trends affecting the reproductive organs of both men and women. Thus, the incidence of testicular cancer has increased quite dramatically in countries with cancer registries, including Scandinavia, the countries around the Baltic Sea, Germany, United Kingdom, United States of America, and New Zealand. Similarly, there has been an increase in the incidence of breast cancer in many countries and the incidence of prostate cancer also appears to have risen. (*European Workshop* 1997, part 1, 6)

While improved reporting may be responsible for some of the increases, problems related to endocrine-disrupting chemicals cannot be excused entirely. Furthermore, according to the Weybridge Report, the apparent decline in male sperm counts in some countries "is likely to be genuine, and not attributable to confounding factors or methodological variables" (*European Workshop* 1997, part 1, 6). The same report cited "insufficient evidence to definitely establish a causal link" between the health effects seen in humans and exposure to chemicals (*European Workshop* 1997, part 1, 6). Concern and further investigation are warranted, however, according to the report (part 1, 14).

POPs had contaminated the environment by the time scientists discovered that many of them mimic the female hormone estrogen, and thereby disrupt reproduction, specially in animals (such as humans) who reside at the top (or end) of their respective food chains. Describing these "endocrine disrupters." Pat Costner observed

The body's hormone system, also known as the endocrine system, is a complex internal chemical messenger system, which regulates vital functions such as our reproductive systems, behavior, and immune systems. In particular, the hormone system controls the development of these vital functions in the unborn child. Hormones are produced by a variety of glands in different parts of the body and released into the blood stream. The hormones bind to special receptors in organs or tissues and cause them to respond in a specific way. Hormones are extremely powerful, having effects at levels of only parts per trillion. But in our bodies, concentrations of hormones are strictly controlled. (Costner 1997)

During 1993 the U.S. federal government's National Institute of Environmental Health Sciences in Bethesda, Maryland, published a report describing the developmental effects of synthetic chemicals on humans and animals that occur in the womb or the ovum. Theo Colborn was the primary author of this report, which considered the following endocrine-disrupting chemicals: 2,4,-D, 2,4,5-T, alachlor, amitrole, atrazine, metribuzin, nitrofen, trifluralin, benomyl, hexachlorobenzene, mancozeb, maneb, metiram-complex, tributyl tin, zinab, ziram, beta-HCH, carbaryl, chlordane, dicofol, dieldrin, DDT and metabolites, endosulfan, heptachlor and heptachlor epoxide, lindane (gamma-HCH), methomyl, methoxychlor, mirex, oxychlordane, parathion, synthetic pyrethroids, toxaphene, transnonachlor, aldicarb, DBCP, cadmium, dioxin (2,3,7,8-TCDD), lead, mercury, PBBs, PCBs, pentachlorophenol (PCP), penta- to nonylphenols, phthalates, and styrenes.

According to this report, "Damage occurs in three key bodily systems: the reproductive system, the endocrine system, and the immune system" (Montague 1993). The endocrine system comprises specialized cells, tissues, and organs that create and secrete hormones, usually into the blood), which then regulate other cells. Hormones thus act as messengers, "sending chemical signals that control the way the entire body grows, is organized, and behaves" (Montague 1993). Colborn, vom Saal, and Soto (1993) describe how endocrine-disrupting chemicals mimic hormones, with one key difference. Natural hormones do their work as messengers,

and then the body disassembles [natural hormones] and removes them from the blood stream. In contrast, when industrial chemicals and pesticides mimic hormones, they do not disappear quickly. They tend to remain in the body for very long periods, doing the work of hormones at times, and in ways, that are inappropriate and destructive. . . . Effects of exposure during development are permanent and irreversible. (378)

Development of the male reproductive tract during fetal life in humans is very sensitive to levels of estrogen. Higher-than-usual estrogen levels may have a role in reducing sperm counts. Chemicals that mimic the effects of estrogen in the body may cause the feminization of many exposed males (Sharpe 1993, 357).

Organochlorines cause specific responses that are usually triggered only by natural estrogens. Estrogenic chemicals also may alter estrogen metabolism by modulating the number of estrogen hormone receptors and their binding affinities. Estrogens play a crucial role in controlling reproduction in women and to a lesser extent in men; they also are involved in fetal development. Therefore, interference by exogenous estrogens can have far-reaching effects on the body (Allsopp, Costner, and Johnston 1995).

Because many disorders of reproductive system have a common origin in fetal life or childhood and most do not become evident until adulthood, the increase in disorders seen today probably originated twenty to forty years ago. The prevalence of such defects in male babies born today, at today's levels of contamination, will therefore not become manifest for another twenty to forty years (Danish EPA 1995).

## FALLING SPERM COUNTS

During the last thirty to fifty years disorders in men related to reproductive organs have risen sharply in several countries. Such disorders include declines in sperm count, according to some studies (e.g., Carlsen et al. 1992) totaling 50 percent during fifty years. The fact that these fifty years (1945–1995) coincide with the advent and worldwide diffusion of synthetic organochlorines may be no coincidence. A study by Elkington (1985) indicated that American men during the 1980s produced half the sperm they did fifty years earlier. Addressing causation, "the research strongly suspects . . . the increased use of herbicides, pesticides, and a chemical flame retardant used in foam mattresses" (Hynes 1989, 22).

Niels Skakkebaek, a pediatrician at the National University Hospital in Copenhagen (where he directs the Department of Growth and Reproduction), discovered the dramatic fall in sperm counts while he was seeking an explanation for unusually high rates of testicular cancer in Danish men. His work led him to study worldwide reports of sperm counts. Skakkebaek studied sixty-one sperm-count reports in twenty-one countries throughout North America, South America, and Asia. Their summary showed a decrease in mean sperm count from 113 million per milliliter in 1940 to 66 million per milliliter in 1990. Skakkebaek (1972) and Sharpe (1993) attributed the decline at least in part to estrogenic organochlorines.

While Skakkebaek's work has been debated, it also has been supported by other researchers. In 1992, for example, E. Carlsen and colleagues (1992) published an analysis of studies of male sperm counts, a summary of studies from various nations with data on almost 15,000 men. Results indicated a large drop in the mean sperm count of 42 percent between 1940 and 1990. This study has been criticized because sperm counts rose slightly after most of the studies documented by Carlsen were conducted. During the 1990s, however, new studies revealed that sperm counts again were generally declining (Auger et al. 1995; Irvine 1994). In addition, a study published in the *British Medical Journal* (Stewart et al. 1996) reported a 24 percent decline in motive (actively swimming) sperm in Scottish men born during the 1950s. The study indicated as well that men born later tended to have lower sperm counts.

Medical researchers' published reports of dramatic declines in sperm counts and increasing sperm abnormalities over the past half century have caused a contentious debate over whether these changes are related to organochlorines. Two of Europe's leading reproductive researchers have hypothesized that increasing exposure to environmental estrogens, which include several POPs, is likely to be responsible not only for lowered sperm counts but also for genital defects, testicular cancer, and other male reproductive abnormalities. Animal studies have also made it clear that humans are currently exposed to levels of dioxins that are roughly equivalent to levels that have caused significant sperm-count declines in male rats exposed in the womb. As researchers probe the cause of the reported human sperm-count declines and other male reproductive problems, POPs stand high on the list of suspects.

## SPERM COUNTS AND AGRICULTURAL CHEMICALS

Researchers during 2002 published findings indicating that semen quality differs significantly between regions of the United States in patterns suggesting that fertile men in more rural areas have lower sperm counts and less vigorous sperm than men in urban areas. The researchers associated environmental factors, such as extensive use of agricultural chemicals, to these differences. Fertile men in Missouri's rural Boone County were found to have a mean sperm count of about 59 million per millimeter, compared to

103 million for men in New York City, 99 million in Minneapolis, and 81 million in Los Angeles. The sperm of men in Boone County were not only less numerous; they also were less vigorous, according to the study ("Rural Men" 2002, A-11; Swan et al. 2002).

Dr. Shanna Swan, an epidemiologist and research professor of Family and Community Medicine at the University of Missouri–Columbia, led a group of researchers who studied 512 couples receiving prenatal care at clinics in Columbia; Minneapolis, Minnesota; Los Angeles, California; and New York, New York. They found that semen quality was equally high in Minneapolis and New York, and slightly lower in Los Angeles. However, men in mid-Missouri had counts and quality that were significantly lower—slightly more than half—than that of men from any of the urban centers. The study appeared in the November 11, 2002, online edition of *Environmental Health Perspectives*, a scientific journal published by the National Institute of Environmental Health Sciences. "We believe that agricultural chemicals could be contributing to this decrease in semen quality," Swan said. "The county in which our Missouri participants lived is quite rural. In 1997, 57 percent of the land was used for farming, compared to 0 to 19 percent for the other three [urban] counties we studied. We are continuing this research and examining the exposure of men to specific chemicals used in farming" ("Chemical Exposure" 2002; Swan et al. 2002). According to an account by the Environment News Service ("Chemical Exposure," 2002), "The only other published study on a comparable semi-rural population analyzed semen quality among men in Iowa City, and also found reduced sperm concentration. Swan and her colleagues are now studying semen quality in Iowa City."

## MALE SEXUAL DYSFUNCTION

In addition to falling sperm counts, measures of several other male sexual dysfunctions rose sharply during the closing decades of the twentieth century. Incidence of testicular cancer in men under age thirty-four has been increasing rapidly in many countries. Recent studies suggest this cancer in young men arises from events early in life or even in the womb, as evidenced by the higher rates of testicular cancer among men with developmental defects such as hypospadias (incomplete masculinization of the male genitals) and undescended testicles.

In the United States, young men fifteen to thirty years of age experienced a 68 percent rise in the rate of testicular cancer between 1972 and 2000 (Moyers 2001). Between 1962 and 1981 the frequency of undescended testicles doubled in England and Wales. The rate of hypospadias in the United States doubled in male infants during the last quarter of the twentieth century (Chilvers, Pike, and Foreman 1984; Jackson, Chilvers, and Pike 1986; Paulozzi, Erickson, and Jackson 1997). Similar increases were reported at the same time in several European countries and Japan. Hypospadias also has been reported when human males were exposed prenatally to antiandrogens such as DDE, a by-product of DDT's breakdown (World Wildlife Fund 2000).

During development of the human fetus, hormones orchestrate key events such as sexual differentiation and the construction of the brain, and so synthetic chemicals that interfere with hormone messages, including each one of the dirty dozen, can disrupt development and cause lifelong damage. In one study of dioxins, reported by the U.S. National Academy of Sciences (2000), a human fetus tested as being 100 times as sensitive as an adult. In the same report, a single low dose of dioxin administered to a pregnant rat at a critical moment in pregnancy did permanent damage to the reproductive systems of her pups, which showed notably diminished male sexual behavior and a sperm-count drop of as much as 40 percent. The dose was very close to the levels of dioxins and related compounds generally reported in people in industrialized regions such as Europe, Japan, and the United States. In female fetuses, the most vulnerable organs are the breasts, fallopian tubes, uterus, cervix, and vagina. In male fetuses, the critical organs are prostate, seminal vesicles (where sperm originates), epididymides (reservoirs for sperm), and testicles. In both sexes, critical organs are the external genitals, the brain, skeleton, thyroid, liver, kidney, and immune system.

Other disorders of the male reproductive system also have increased. Testicular germ-cell cancer is now the most common malignancy among young men in many industrialized countries. The Danish Cancer Registry has collected reliable data since 1943 on the incidence of testicular germ-cell cancer showing a three- to fourfold increase between the 1940s and 1980s. Studies based on data from cancer registries in other countries, including the United Kingdom, the Scandinavian and Baltic countries, Australia, New Zealand, and the United States also have reported significant increases in the incidence of testicular cancer.

## THE YUCHENG INCIDENT AND DECLINING SPERM COUNTS

A large number of Taiwanese were poisoned during 1978 and 1979 in the Yucheng Incident as they consumed PCB- and dioxin-contaminated rice that had been cooked with polluted oil. The rice oil contained 100 ppm PCBs and 0.1 ppm PCDFs [polychlorinated dibenzofurans]. Two studies describing the sexual development of children born to women who were pregnant at that time (Guo, Chen, et al. 1994; Guo, Lai, et al. 1993) revealed that penis sizes of boys born seven to twelve years after the poisoning were significantly shorter than the penises of matched control children. The penises were probably stunted because of *in utero* exposure to estrogenic PCBs (Holloway 1994, 25). Roughly 2,000 people consumed the contaminated oil. The children consumed no contaminated oil themselves; they were exposed before birth to PCBs that were carried by their mothers' milk (Montague 1994; Rogan et al. 1988, 334).

The Yucheng or "oil disease" children were followed medically for several years. When 115 Yucheng children were examined in 1985, they were less developed than a control group of children on thirty-two of thirty-three measures. Compared to controls, they were delayed in the age that they performed tasks such as saying phrases and sentences, turning pages, carrying out requests, pointing to body parts, holding pencils, and catching a ball. The Yucheng children also had a variety of physical defects at birth, including dark-colored heads, faces, and genitals, and abnormal nails that were often dark and ridged, split, or folded (Kolata 1988; Montague 1994). These children provided the first direct evidence that PCBs produce birth defects—that they are, in medical terminology, teratogenic.

## INFERTILITY AND DBCP

Other organochlorine pesticides, such as dibromochloropropane (DBCP), have caused infertility in men who have been exposed to the compound at their work sites (Whorton and Foliart 1983). DBCP was used as a soil fumigant beginning in the mid-1950s; its use peaked during the mid-1970s, before disclosures of infertility provoked the U.S. EPA to ban its manufacture and use in 1979.

The first indication that DBCP caused infertility in men arose in 1977, when a group of five male workers at a California pesticide formulation plant sought medical advice after talking among themselves about their shared inability to father children. All five were found to have low sperm counts. Further investigation at the plant produced a very strong relationship between the duration of DBCP exposure and sperm-count decline. According to one observer, "The exact mechanisms by which DBCP exerts adverse effects on the male reproductive system are unknown, but it is clear that it damages and destroys the germ cells that produce sperm. This results in a reduction in the numbers and motility of sperm that causes infertility. Sperm counts have recovered in some of the men affected by DBCP, but others remain sterile" (Allsopp, Costner, and Johnston 1995).

Large amounts of DBCP were used in the Atlantic banana-growing regions of Costa Rica between 1971 and 1978. Workers were typically exposed to the chemical for long periods of time. By mid-1990, approximately 1,500 male workers from these banana plantations were diagnosed as infertile. Physicians on the scene estimated that about a thousand similar cases had not been reported. Between 60 and 70 percent of all sterile victims were azoospermic (having no sperm to count) and the rest are oligospermic (having low sperm counts of less than 20 million per milliliter). Since virility is often accorded a high social value among men in Latin America, its loss due to DBCP toxicity provoked other psychological and social effects, including depression (in more than half the patients), impotence, and an increasing number of divorces attributable, at least in part, to DBCP-induced sterility (Thrupp 1991).

Dioxins have displayed adverse effects on male reproductive systems in animals and humans from birth to death. Experiments on rats and other laboratory animals have shown that exposure of adult males to high levels of dioxins produces changes in the levels of male reproductive hormones, including a reduction in testosterone, decreased sperm formation (and fertility), as well as reduced weights of the testis (Peterson, Theobald, and Kimmel 1993). During the Vietnam War, dioxin-contaminated Agent Orange was sprayed liberally in Southeast Asia by U.S. armed forces (see chapter 1). Members of the U.S. armed forces who sprayed Agent Orange often were contaminated with dioxins. Some of these men's testes shrank after exposure, an effect that has been linked by Wolfe, Michalek, and Miner (1994) to dioxin contamination.

## SYNTHETIC ESTROGEN AND BREAST CANCER

The incidence of breast cancer has steadily risen worldwide. By the 1990s, it was the leading cause of death from cancer for women in the United States (El-Bayoumy 1992). Breast cancer mortality has increased at an estimated rate of 1 percent per year since the 1940s in the United States, even allowing for increased detection rates by mammography. Between 1982 and 1986, breast-cancer mortality rose by 4 percent a year. Increases have occurred in all age groups (Harris et al. 1992).

Because synthetic estrogenic chemicals in the environment can mimic natural endogenous estrogens or interfere with their metabolism, it has been hypothesized that such chemicals may increase the risk of breast cancer (Davis et al. 1993). A number of organochlorine compounds (DDT and symmetrical triazine herbicides such as simazine chlortriazine and Atrazine) have been shown to induce and promote breast cancer in laboratory animals (Stevens et al. 1994; Wetzel et al. 1994). Increased concentrations of estrogens during pregnancy also may increase the risk of breast cancer in daughters (Trichopoulos 1990). According to expert testimony, "There is considerable evidence that the total lifetime exposure to estrogen influences the likelihood of developing breast cancer. High serum or urine levels of estrogen, early onset of menstruation, delayed menopause, and delayed first-child bearing are all risk factors for breast cancer" (Schettler et al. 1999, 160). Risk factors also include radiation exposure and alcohol consumption.

The rising level of synthetic estrogens in the environment may not be the *single* cause of rising breast-cancer rates during the last half of the twentieth century, but one factor among many that influence lifetime exposure. Commented Schettler and colleagues (1999), "A considerable amount of interest and research is focussed on . . . organochlorine compounds, solvents, metals, and polycyclic aromatic hydrocarbons, which are products of combustion spread widely throughout the environment. Breast milk contains a large number of these contaminants in complex mixtures" (Schettler et al. 1999, 160).

Because of breast cancer's complex origins, scientists have had trouble tracing singular causes. Given that caveat, breast-cancer rates have risen substantially during the time that POPs have become pervasive in the environment. During the middle 1940s, one in twenty-two women was afflicted with breast cancer at some point in her lifetime. By the end of the 1980s, that ratio was roughly one in eight (Cadbury 1997, 220). Some of this rise may be due to increased detection and treatment of the disease. The rest of the increase defies singular or simple explanation.

A link between estrogenic chemicals and cancer was suspected as early as the 1930s by E. C. Dodds and colleagues (Dodds, Goldberg, and Lawson 1938; Dodds et al. 1938; Dodds and Lawson 1938). A. P. Høyer and colleagues (1998) showed that women, once afflicted by breast cancer, do not survive as long if they have relatively high levels of dieldrin in their blood.

Dieldrin had a significant adverse effect on overall survival and breast cancer specific survival . . . These findings suggest that past exposure to estrogenic organochlorines such as dieldrin may not only affect the risk of developing breast cancer but also the survival. After diagnosis for breast cancer, women with highest dieldrin levels survive the shortest time, on average. A high serum dieldrin concentration was consistently related to a subsequent poorer survival. . . . It is therefore also possible that exposure to organochlorines somehow induces the aggressiveness of the tumor. (Høyer et al. 2000, 323)

## SEX RATIOS AFTER THE SEVESO, ITALY, DIOXIN SPILL

The first scientific evidence that low-level dioxin pollution has a direct effect on reproduction was revealed by Italian researchers who studied the population of Seveso, in northern Italy, the scene of a devastating explosion at a chemical factory in 1976 (see chapter 1). The explosion released a large quantity of dioxins into the atmosphere. The researchers found that babies born of women living near the scene of the disaster had a sharply skewed sex ratio, with more girls than boys. The effects of dioxin exposure continued to shape the reproductive history of the area a full quarter-century after the accident.

Paolo Mocarelli and colleagues in the department of laboratory medicine at Desio Hospital, in Milan, found that men with the highest dioxin levels were least likely to father boys. The sex ratio of children for parents aged under nineteen at the time of the accident was 62 boys born for every 100 girls, compared with a usual sex ratio of 106 boys to 100 girls, according to a study that was published in the *Lancet* (Mocarelli et al. 2000). The families of women who were exposed to high levels of dioxin and married men from outside the area were unaffected. The families of exposed men who married women from outside the area had the skewed sex ratio, however.

This effect was evident even at very low levels of 20 nanograms per kilogram of body weight, only twenty times the estimated average concentration of dioxin in humans in industrialized countries. According to Mocarelli and colleagues: "The observed effects . . . started at concentrations of less than 20 nanograms per kilogram of bodyweight. This could have important public health implications" (1858). Specifically how dioxin affects the male reproductive system is unclear. Scientists have reported a decreased proportion of male births in Denmark, the Netherlands, the United States, and Canada and in certain occupational groups, including sawmill workers and in those exposed to air pollution from incinerators.

Professor Mocarelli said that assessing a safe level of dioxins is complicated. "Relatively low levels of dioxin are having an effect on males. We have shown for the first time that the human male reproductive system is very sensitive to dioxin" (Laurance 2000). Mocarelli also commented, "Dioxin contamination is a world-wide problem. These data will assist health authorities to better define risk assessment for these toxic molecules. In fact the lowering sex ratio in human beings has been directly linked for the first time to male exposure to an environmental pollutant and at relatively low doses. This effect can be tentatively interpreted as a result of dioxin endocrine disruption of the male reproductive system" (*Lancet* Press Release 2000).

## SYNTHETIC ESTROGEN (DES)

Between 1945 and 1971, many women were treated with a synthetic estrogen called diethylstilbestrol (DES), which was given to at least 5 million women to prevent miscarriages and pregnancy complications. The chemical was banned in 1971 after it was linked to increases in a previously rare form of vaginal cancer. Women who were exposed to DES *in utero* also have suffered an increased incidence of cervical cancer in adulthood, as many of them also experienced increased fertility problems (Hines 1992). Theo Colborn and colleagues (1993) assert that "DES-exposed humans . . . serve as a model for exposure during early life to any estrogenic chemical" (378).

The FDA approved DES for hormone therapy and relief of menopausal disorders in 1941; during 1947, it was approved to promote the growth of chickens and in 1949 for the prevention of miscarriages. The sales push for DES was undertaken despite evidence that it was good for nothing. By 1952 at least four independent studies had reported that "women treated with DES for threatened miscarriages did no better than those treated with alternatives such as bed rest or sedatives" (Colborn, Dumanoski, and Myers 1996, 54). In 1954 DES was approved as a growth enhancer for sheep and cattle.

By the late 1950s, DES was being promoted for a wide range of maladies, as a miracle drug. During 1957 the *Journal of Obstetrics and Gynecology* published advertising from the Grant Chemical Company, a maker of DES, promoting it for "ALL pregnancies" to produce "bigger and stronger babies" (Colborn, Dumanoski, and Myers 1996, 48). Doctors prescribed DES to suppress milk production after childbirth, to alleviate hot flashes and other symptoms of menopause, and to cure acne, prostate cancer, and "even to stunt the growth [of] teenage girls who were becoming unfashionably tall" (Colborn, Dumanoski, and Myers 1996, 48).

The FDA suspended the use of DES in poultry feed in 1958 after it was shown to be carcinogenic in experimental animals. It was thirteen years after that, however, before its use on humans was outlawed in 1972, the same year the U.S. EPA banned DDT. By the time DES was banned in humans, its carcinogenic properties were very painfully obvious.

The ban on DES use in women was provoked by Arthur Herbst, who discovered a relationship between vaginal cancer and DES during his tenure at Harvard Medical School and Massachusetts General Hospital. Commented Krimsky, "Men exposed to DES during gestation were more likely to exhibit abnormalities in sex organs, reduced sperm count, and lower quality of semen compared with controls" (Krimsky 1999, 10).

In the world of toxicology, study of the DES debacle has provided some important lessons. Experience with DES, for example, demonstrated that synthetic chemicals could cross the placenta and disrupt the extremely delicate balance of hormonal development. The DES experience also illustrated that such contamination could manifest itself many years after the initial shock of exposure in the womb.

Further investigation of the effects of DES revealed that *in utero* exposure resulted in a much greater incidence of cryptorchidism and hypospadias in boys. In addition, when they reached adulthood, many of the same men had reduced sperm counts. The same outcome was achieved in the male offspring of mice and rats to which DES had been administered (Sharpe 1993). The number of people exposed to DES (mothers and offspring) between 1950 and 1971 has been estimated at 5 to 10 million (Schettler et al. 1999, 153).

My sister, Linda Carol Edgar, was one of them. I was born January 30, 1950, six weeks early, and severely underweight, barely surviving. When my mother became pregnant with my sister Linda, she took DES to prevent another premature birth. Nineteen months after my birth, Linda came into the world at weight and on time, as promised. Twenty years later, she lost an ovary to that previously rare form of cervical cancer that had suddenly become so pervasive among women born to mothers who had taken DES to facilitate pregnancy.

## HORMONE MIMICS AND PROSTATE PROBLEMS

Prenatal exposure to hormone-mimicking substances also "may be exacerbating the most common problem afflicting aging males: painfully enlarged prostate glands that make urination difficult and often require surgery" (Colborn, Dumanoski, and Myers 1996, 179–80). Prostate cancer rates rose 126 percent in the United States between 1973 and 1991, according to the National Cancer Institute. Low doses of estrogen over a long period of time have been found (by Shuk Meri Ho of Tufts University) to induce prostate cancer in mice.

## ALKYLPHENOLS AND PHTHALATES

Alkylphenols and phthalates are weakly estrogenic and have been linked to decreased testicular size, reduced sperm counts, and male feminization in some animal studies (Schettler et al. 1999, 179). About 1 billion pounds of these compounds are produced in the United States each year. They are industrial chemicals that are widely used in detergents, paints, pesticides, plastics (including food wraps), and many other consumer products. Alkylphenols may be found in drinking water and leaching from plastics used in food processing and wrapping, as well as construction materials and children's teething rings.

Two dozen varieties of phthalates are among the most abundant of manufactured chemicals; they also are used in construction, in automotive and medical industries, and in a wide variety of consumer items, including many toys (the ubiquitous Barbie doll among them), some clothing, and packaging. Most plastic wraps, beverage containers, and linings of metal cans contain phthalates. The most common use of phthalates is to plasticize (impart softness) to polyvinyl chloride (PVC). Phthalates are sometimes mildly estrogenic and have been identified as testicular and ovarian toxicants in animal studies (Schettler et al. 1999, 181).

"Recent laboratory research in animals has linked phthalates to damage to the liver, kidney, and testicles, as well as to miscarriage, birth defects, and reduced fertility," according to one researcher (McGinn 2000, 34). Incineration of phthalates produces dioxins also.

## CONCERNS ABOUT PVC PLASTICS

During the 1990s, the market for PVCs was the fastest-growing chlorine-related market in the world. Products manufactured from PVCs had been incorporated into a wide variety of industrial processes and products that most people use in their everyday lives, from toothpaste tubes to children's toys. If a plastic is flexible, it probably contains PVC.

Dioxins are produced throughout the lifecycle of PVC plastic (Thornton 1997). The manufacture of PVC begins when chlorine gas is produced by the electrolysis of brine, during which dioxin is formed. Chlorine is then combined with ethylene to produce ethylene dichloride (EDC). During this process large quantities of dioxins are formed, some of which are released into the air and water. Nearly 40 percent of all the chlorine produced by the chemical industry is used in the manufacture of PVC (Thornton 1997). Although no direct comparative data exists, PVC production is probably the single largest source of dioxins in the environment.

Joe Thornton has written, under the aegis of Greenpeace, that "Samples taken downstream from EDC manufacturers in the U.S. and Europe indicate significant contamination of sediment and the food chain in the vicinity of these plants" (Thornton 1997). Greenpeace analyses at U.S. chemical facilities have indicated that wastes from the creation of EDC are among the most intensely dioxinated wastes anywhere. These chlorine-rich wastes often are incinerated, producing and releasing dioxins. Once it has been created, EDC is then converted into vinyl chloride monomer (VCM), which is polymerized, formulated, and formed into a final PVC-containing product.

Plastics containing PVC thus create dioxins as they are created. These plastics also create dioxins when they are incinerated. Dioxins also may be produced in other ways. When buildings or vehicles are consumed by fire, dioxins usually are produced. More typically, PVC plastics are incinerated or smelted in heat-reactive processes that add to the ecosphere's load of dioxins. More than 1 billion pounds of PVCs may be burned in U.S. trash and medical-waste incinerators and in accidental structural fires each year (Thornton 1997). Thornton adds that "The unknown quantities of PVC burned in industrial and warehouse fires, automobile fires, metals smelters, and wood combustion add to this total" (Thornton 1997).

Furthermore, according to Thornton, while production of many chlorinated compounds is declining and some are holding steady, worldwide production of PVC and its feedstocks "is rapidly growing, both for use in the U.S. and for export, primarily to developing nations" (Thornton 1997).

PVC plastic was first marketed in 1936. A 1966 Goodrich Corporation memo admitted that "there is no question but that skin lesions, absorption of bone of the terminal joints of the hands and circulatory changes can occur in workers associated with the polymerization of PVC" (Kurtz 2001, C1). "In other words, they knew vinyl chloride could cause the bones in the hands of their workers to dissolve," Bill Moyers told a Public Broadcasting System audience in March 2001 (Kurtz 2001, C1). Another document, a 1973 Ethyl Corporation memo, says that the results of rat tests "certainly indicate a positive carcinogenic effect" (Kurtz 2001, C1).

During the early 1970s, an Italian scientist, P. L. Viola (1971, 516), exposed laboratory rats to vinyl chloride and diagnosed unusually high rates of cancer. As Viola lowered exposure levels in his tests, cancer persisted. A year later, another Italian researcher, Cesare Maltoni, found evidence of a rare liver cancer, angiosarcoma (Moyers 2001). In Maltoni's studies, cancer appeared in rats exposed to levels of vinyl chloride that were common on factory floors in the United States at the time. According to the report by Bill Moyers (2001), a Union Carbide executive reported to corporate headquarters that if a letter admitting knowledge of Maltoni's work ever became public, it could "be construed as evidence of an illegal conspiracy by industry . . . if the information were not made public or at least made available to the government."

During the 1970s, facing mounting evidence of PVC's toxicity, the U.S. federal government, rejecting contrary advice from the industry, ordered workplace exposure to vinyl chloride reduced to 1 ppm. During 1974 B.F. Goodrich announced that four workers at its Louisville, Kentucky, vinyl chloride plant had died from angiosarcoma, a previously rare liver cancer linked to PVC manufacture, as Maltoni had found in his study of rats. After the four deaths, 270 employees at the plant were tested. Blood abnormalities were detected in 55 of them.

Moyers (2001) interviewed the wife of Dan Ross, a worker for Conoco, and she described how he returned home one day from the PVC plant. He took off his boots, and his wife looked at his feet. She recalled

The whole top of 'em were burned. Now, he had on safety boots, steel-toed, and the whole top of his feet were red where the chemicals had gone through his boots, through his socks, under his feet, and burned them, both feet.

Later, cancer was discovered in Ross's brain. Ross's wife recalled how he coped with his rapid physical decline:

You start watching him die one piece at a time, you know. It's like, okay, he's blind today, but he can still hear, he can still swallow if I put something in his mouth. But he lost the use of one of his arms, and then next day it would be the other arm, the next day it would be one leg. And then he couldn't hear anymore. The hardest part was when he couldn't speak anymore. (Moyers 2001)

On October 9, 1990, twenty-three years to the day after he had started working at Conoco, Dan Ross died, at age forty-six. "In the last words he was able to speak, Dan Ross told his wife, "'Mama, they killed me'" (Moyers 2001).

The rising volume of PVC production is also adding to the environment's overload of PCB pollution. While the direct production of PCBs was banned by the United States during 1977, this persistent pollutant is still released into the ecosphere through the generation of wastes, one of which is the manufacture of PVCs. In 1990 Dow Chemical analyzed some of its chlorinated wastes and found that they contained 302 ppm PCBs.

Manufacture of vinyl chloride monomer has caused extensive dioxin contamination of Rotterdam harbor in the Netherlands. In Venice, Greenpeace analyzed sediment from the Porto Marghera that showed dioxin contamination of the lagoon by the Enichem plant, where vinyl chloride monomer is manufactured. In Germany, the Environmental Ministry of Lower Saxony found extremely high levels of dioxins in sludges from the wastewater treatment plant at Wilhelmshaven. Dioxins also were found in a dump where these sludges were disposed (Costner 1997).

PVC-coated wallpaper and wood (often as furniture) often are burned by individuals in their backyards, or by small companies in inadequate furnaces that are not suited for burning such hazardous wastes. In addition, in many parts of the world, plastic cable scrap containing PVC is burned in the open air. For example, in 1994 Greenpeace discovered that imported PVC cables are recycled in the slums of Jakarta, Indonesia, simply by burning the PVC off the cables in big steel drums in people's backyards (Costner 1997).

## MORE POTENTIAL PROBLEMS WITH PHTHALATES

Hair sprays, perfumes, and other brand-name cosmetics contain toxic chemical phthalates that may be absorbed into the human body, according to the first independent tests for these chemicals in over-the-counter products. According to a report by Cat Lazaroff for the Environment News Service, "Christian Dior's Poison perfume, Arrid Extra Extra Dry deodorant and Aqua Net hair spray are among many of the beauty and personal-care products that contain one or more of the dangerous chemicals known as phthalates." Lazaroff's report was drawn from a report titled *Not Too Pretty*, released July 10, 2002, by the Environmental Working Group,

Coming Clean, and Health Care without Harm. The groups con-
tracted with a major national laboratory to test seventy-two name-
brand, off-the-shelf beauty products for the presence of phthalates,
a large family of industrial chemicals linked to birth defects in the
male reproductive system. The lab found phthalates in fifty-two of
the seventy-two, or 72 percent, of the products tested. Only one of
the products listed phthalates on the label (Lazaroff 2002).

Phthalates are used as a plastic softener and solvent in several
consumer products. In cosmetics and other beauty products,
phthalates help make nail polish chip-resistant and extend the life
of perfumes' fragrance. According to this report, these chemicals
can be absorbed through the skin, inhaled as fumes, ingested when
they contaminate food or when children bite or suck on toys, and
inadvertently administered to patients from plastic medical devices
containing PVC. Numerous animal studies have demonstrated that
phthalates can damage the liver, kidneys, lungs and reproductive
system, particularly the developing testes, the groups sponsoring
the study asserted (Lazaroff 2002).

According to the Environment News Service report, One Centers
for Disease Control (C.D.C.) study found that 5 per cent of women
of reproductive age—an estimated two million women—may be getting up to 20 times more of the phthalate D.B.P. than the average
person in the population. The highest exposures for women of child-
bearing age were above the federal safety standard, which may create a risk of reproductive birth defects, based on animal studies
considered relevant to humans (Lazaroff 2002).

In reply, the Cosmetic, Toiletry and Fragrance Association (CTFA),
an industry group, said that the use of phthalates in cosmetics and
personal care products is supported by "an extensive body of scientific research and data that confirms safety. The Food and Drug
Administration (F.D.A.), the Environmental Protection Agency
(E.P.A.), Health Canada and other scientific bodies. In Europe,
North America, and Japan have examined phthalates and allow
their continued use," the CTFA said (Lazaroff 2002). "Phthalates
were also reviewed by the Cosmetic Ingredient Review (C.I.R.), an
independent body that reviews the safety of ingredients used in cosmetics," the group added. "C.I.R. found them to be safe for use in
cosmetics in 1985" (Lazaroff 2002). The CIR's expert panel voted
to begin a re-review of phthalates. The CTFA acknowledged but
stressed that the re-review "does not suggest that a previous con-

clusion will be changed, only that there is sufficient new information that should be evaluated" (Lazaroff 2002).

"The testing done for *Not Too Pretty* covers less than one per cent of the beauty products sold in drug and discount stores across the United States, but it appears to be the most comprehensive testing ever made available to American consumers," said Charlotte Brody, executive director of Health Care without Harm. "Because of lax F.D.A. labeling rules, we cannot know how many more beauty products contain unlabeled quantities of phthalates" (Lazaroff 2002).

According to the Environment News Service, the limited testing done for *Not Too Pretty* revealed that the same companies that produce phthalate-laden beauty products also make products free of phthalates. For example, Unilever makes hair sprays with phthalates—Aqua Net and Salon Selectives—and without phthalates—Thermasilk and Suave. L'Oreal markets Jet Set nail polish without DBP but puts the phthalate in its Maybelline brand. Procter and Gamble sells Secret Sheer Dry deodorant with phthalates and Secret Platinum Protection Ambition Scent without phthalates. Louis Vuitton has taken phthalates out of its Urban Decay nail polish but still has these chemicals in Christian Dior nail polish and the fragrance Poison (Lazaroff 2002).

Samuel S. Epstein, chairman of the Cancer Prevention Coalition, and professor emeritus, Environmental Medicine, University of Illinois School of Public Health, Chicago, wrote in a commentary for the Environment News Service, "With rare exceptions such as children's bubble baths, the Food and Drug Administration has never required the industry to label its products with any warning of well-documented risks, particularly reproductive and cancer; nor has the F.D.A. banned the sale of unsafe products to an unsuspecting public, although so explicitly authorized by the 1938 Food, Drug and Cosmetics Act" (Epstein 2002).

The Environmental Health Network of California during mid-2002 petitioned the U.S. Food and Drug Administration asking for warning labels on cosmetics to identify allergens and hazardous substances contained in the hair spray, deodorant, nail polish, and perfume. Epstein said that the labels are especially important for phthalates in perfumes and fragrances, to which about 12 percent of the population are sensitive ("Labeling Cosmetics" 2002).

The fragrance industry uses phthalates to make scents evaporate more slowly; "Phthalates in fragrances make the scent last longer," the American Chemistry Council wrote July 10 in "Phthalates and Your Health" ("Labeling Cosmetics" 2002). Besides some twenty allergens, cosmetics and toiletries contain other numerous hazardous ingredients, says Epstein. "These include about 100 carcinogens and 15 endocrine hormonal disrupters" ("Labeling Cosmetics" 2002). As a variety of products is used by a single person, complicating factors come into play that Epstein calls "hidden" carcinogens. "These include those contaminating non-carcinogenic ingredients, those formed in the product or skin by the breakdown of non-carcinogenic ingredients, and those formed by the chemical interaction between ingredients or contaminants," he says ("Labeling Cosmetics" 2002).

## EUROPEAN PARLIAMENT VOTES TO ELIMINATE PVCS

Concern regarding estrogenic properties of PVCs reached a level in Europe that provoked, on April 3, 2001, a European Parliament vote to eliminate polyvinyl chloride plastics. "The European Parliament has recognized the dangers associated with PVC production, use and disposal and voted in the interests of the environment and public health. This is an important step towards effective action against the many hazards of PVC plastic and the need to use safer materials. Evidence that PVC harms the environment and human health is overwhelming and, as today's vote reflects, there is now only one way forward: PVC has got to go," said Greenpeace campaigner Maureen Penjueli ("European Parliament" 2001).

The European Parliament voted in favor of introducing a substitution policy, starting with replacement of soft PVC. The same European union vote also called for a ban of lead additives in PVC and for compulsory marking of PVC products, as well as separate collection of PVC waste. Moreover, the parliament said that incineration and landfills are unsustainable options for PVC disposal. Combustion of PVC leads to hazardous emissions in the atmosphere, such as dioxin, and produces toxic ash that has to be landfilled.

"Concern over the hazards of PVC plastic is widespread. Local communities, health groups, consumer groups and industry have all been calling on the E.U. to phase out this dangerous plastic," said Penjueli. "It's heartening that finally their concerns have been validated and their voices heard. It is now vital that the European

Commission proposes a directive to act on today's decisions," she added.

## TOXIC BARBIE?

Information presented during late August 2000 at an American Chemical Society meeting in Washington, D.C., suggested that some vintage toys (including older Barbie dolls) may ooze PVC as they age, exposing children to endocrine-disrupting chemicals. Yvonne Shashoua, a conservation scientist and chemist with the National Museum of Denmark in Copenhagen, told the meeting that some old Barbie dolls manufactured in the early years after the plastic princess's first release in 1959 contain PVC (Schwanke and Yoder 2000).

Deteriorating Barbie dolls may become sticky, with heat and sunlight accelerating the process. Shashoua says that use of the troublesome substance has been generally banned, and a new formula now used in PVC products does not pose a known health risk. (That is to say, no health risk for which the replacement substances have been tested, which is not a guarantee of safety, given the way research follows use of many chemicals.)

## ESTROGEN MIMICRY AND HEATED PLASTIC

Even microwaved plastic may mimic estrogen. David Feldman of the Stanford University Medical School, who, with colleagues, was seeking the evolutionary origins of steroids (the hormone) found unexplained estrogen mimics in their laboratory samples. They tried and failed to purge the substances from their equipment, even in otherwise sterile samples. Feldman and colleagues learned eventually that estrogen mimics were leaching from the polycarbonate flasks used to sterilize water for experiments, as they detected bisphenol A, a synthetic estrogen, leaching from their plastic lab equipment.

Bisphenol A is similar in molecular structure to DES and also acts estrogenically, but with one-thousandth or two-thousandths the potency. Leaching generally takes place when the plastic is heated. Some plastics (usually those that are more rigid to the touch) do not leach when heated, but softer ones do. If the plastic interacts with the food when it is heated, it may be leaching synthetic estrogens.

This is the same problem as with old Barbie dolls made of PVC, which does leach estrogen mimics when it is heated (Feldman et al. 1984).

Bisphenol A also has been detected in the plastic linings inside many metal food cans (the plastic lining is meant to keep the food from mixing with residues from the metal). Nicholas Olea found that many of these cans with epoxy-based linings "contained Bisphenol A in sufficient quantities that they could make breast cancer cells divide" (Cadbury 1997, 150; Brotons et al. 1995). While natural estrogens are eliminated from the body by natural processes, the synthetics "have the ability to remain in the body for months or years, building up a reservoir in the fat stores" (Cadbury 1997, 171).

## RESEARCH NEEDED ON ESTROGEN MIMICS

In August 2000 the National Research Council of the U.S. National Academy of Sciences made public its long-awaited report, *Hormonally Active Agents in the Environment.* The report cited much of the research by Theo Colborn and others and stressed that hormone disruption is a serious scientific issue deserving of funded research. The potential of endocrine disruption was one of the major elements compelling representatives from about one hundred twenty countries to finalize a ban, in Stockholm, of the dirty dozen organochlorines in December 2000. The detected range of effects keeps expanding, over time, as the amount of any given chemical thought to pose a threat keeps dropping.

A National Toxicology Program coordinated by the National Institute of Environmental Health Sciences issued a report that brought back something of a hung jury on the question of low-dosage endocrine disrupters: "Several studies provide credible evidence for low-dose effects of bisphenol A; these include increased prostate weight in male mice at six months of age" (National Toxicology Program 2000, iii). However, other studies using other rats in other settings found no evidence. The panel concluded that while there is "credible evidence that low doses of BPA [bisphenol A] can cause effects on specific endpoints. . . . [but] the panel is not persuaded that a low-dose effect of BPA has been conclusively established as a general or reproducible finding" (National Toxicology Program 2000, iv; see also vom Saal et al. 1997). The same report found that low-dose effects "were clearly demonstrated for estradiol and several other estrogenic compounds" (National Toxicology Program 2000, iv).

# REFERENCES

Allsopp, Michelle, Pat Costner, and Paul Johnston. 1995. *Body of Evidence: The Effects of Chlorine on Human Health.* London: Greenpeace International.

Auger, J., J. M. Kuntsmann, F. Czyglik, and P. Jouannet. 1995. "Decline in Semen Quality among Fertile Men in Paris during the Past 20 Years." *New England Journal of Medicine* 332, no. 5: 281–85.

Brotons, J. A., M. F. Olea-Serrano, M. Villalobos, and N. Olea. 1995. "Xeno-estrogens Released from Lacquer Coatings in Food Cans." *Environmental Health Perspectives* 102: 608–12.

Cadbury, Deborah. 1997. *Altering Eden: The Feminization of Nature.* New York: St. Martin's Press.

Carlsen, E., A. Giwercman, N. Keiding, and N. E. Skakkebaek. 1992. "Evidence for Decreasing Quality of Semen during the Past 50 Years." *British Medical Journal* 305: 609–13.

"Chemical Exposure May Reduce Sperm Quality." 2002. Environment News Service. November 13. http://ens-news.com/ens/nov2002/2002-11-13-09.asp#anchor7.

Chilvers, C., M. C. Pike, and D. Foreman. 1984. "Apparent Doubling of Frequency of Undescended Testicles in England and Wales, 1962–1981." *Lancet* 2 (8398): 330–32.

Colborn, T., and C. Clement. 1992. "Statement from the Work Session on Chemically-Induced Alterations in Sexual and Functional Development: The Wildlife/Human Connection." In *Advances in Modern Environmental Toxicology,* edited by T. Colborn and C. Clement. Vol. 21. Princeton, N.J.: Princeton Scientific Publishing.

Colborn, T., D. Dumanoski, and J. P. Myers. 1996. *Our Stolen Future: Are We Threatening Our Fertility, Intelligence and Survival? A Scientific Detective Story.* New York: Penguin.

Colborn, Theo, Frederick S. vom Saal, and Ana M. Soto. 1993. "Developmental Effects of Endocrine-Disrupting Chemicals in Wildlife and Humans." *Environmental Health Perspectives* 101, no. 5 (October): 378–84.

Cook, J. W., and E. C. Dodds. 1933 "Sex Hormones and Cancer-Producing Compounds." *Nature* (February): 205–6.

Costner, Pat. 1997. "The Burning Question—Chlorine and Dioxin. Taking Back Our Stolen Future: Hormone Disruption and PVC Plastic." Greenpeace USA. April. http://www.greenpeace.org/~toxics/reports/tbosf/tbosf.html#Introduction.

Danish Environmental Protection Agency. 1995. *Male Reproductive Health and Environmental Chemicals with Oestrogenic Effects.* Copenhagen: Danish Environmental Protection Agency.

Davis, D. L., H. L. Bradlow, M. Wolff, T. Woodruff, D. G. Hoel, and H. Anton-Culver. 1993. "Medical Hypothesis: Xenooestrogens As Preventable

Causes of Breast Cancer." *Environmental Health Perspectives* 101, no. 5: 372–77.

Dieckmann, W. J., M. D. Davis, and R. E. Pottinger. 1953. "Does the Administration of DES during Pregnancy Have Any Therapeutic Value?" *American Journal of Obstetrics and Gynecology* 66:1062–81.

Dodds, E. C., L. Goldberg, and W. Lawson. 1938. "Oestrogenic Activity of Esters of Diethyl Stilboestrol." *Nature* (1938): 211–12.

Dodds, E. C., L. Goldberg, W. Lawson, and R. Robinson. 1938. "Oestrogenic Activity of Alkylated Stilboestrols." *Nature* (1938a): 247–49.

Dodds, E. C., and W. Lawson. 1938. "Molecular Structure in Relation to Oestrogenic Activity: Compounds without a Phenanthrene Nucleus." *Proceedings of the Royal Society of London* 125, suppl. B: 222–32 (1938c).

Eaton, S. B., M. C. Pike, R. V. Short, N. C. Lee, J. Trussell, R. A. Hatcher, J. W. Wood, C. M. Worthman, N. G. Blurton Jones, M. J. Konner, K. R. Hill, R. Bailey, and A. M. Hurtado. 1994. "Women's Reproductive Cancers in Evolutionary Context." *Quarterly Review of Biology* 69, no. 3: 353–67.

El-Bayoumy, K. 1992. "Environmental Carcinogens That May Be Involved in Human Breast Cancer Etiology." *Chemical Research in Toxicology* 5, no. 5: 585–90.

Elkington, J. 1985. *The Poisoned Womb: Human Reproduction in a Polluted World.* London: Harmondsworth.

Epstein, Samuel S. 2002. "Beware Carcinogens, Phthalates in Cosmetics." Environmental News Service. July 15. http://ens-news.com/ens/jul2002/2002-07-15e.asp.

European Commission. 1997. *European Workshop on the Impact of Endocrine Disrupters on Human Health and Wildlife: Report of Proceedings, December 2–4, 1996, Weybridge, U.K.* Copenhagen, Denmark: European Commission. [Report EUR 17549.]

"European Parliament Votes for Substitution of PVC Plastic." 2001. Press release, Greenpeace London. April 3. www.greenpeace.org.

Feldman, D., L. G. Tokes, P. A. Stathis, and D. Harvey. 1984. "Identification of 17B-oestradiol As the Estrogenic Substance in Saccharmyces Cerevisae." *Proceedings of the National Academy of Sciences* 81: 4722–28.

Guo, Y. L., Y. C. Chen, M. L. Yu, and C. H. Chen. 1994. "Early Development of Yu-Cheng Children Born Seven to Twelve Years after the Taiwan PCB Outbreak." *Chemosphere* 29, nos. 9–11: 2395–404.

Guo, Y. L., T. J. Lai, S. H. Ju, Y. C. Chen, and C. C. Hsu. 1993. "Sexual Development and Biological Findings in Yucheng Children." *Organohalogen Compounds* 14: 235–38.

Harris, J. R., M. E. Lippman, U. Veronesi, and W. Willett. 1992. "Breast Cancer (First of Three Parts)." *New England Journal of Medicine* 327, no. 5: 319–28.

Hines, M. 1992. "Surrounded by Oestrogens? Considerations for Neuro-behavioural Development in Human Beings." In *Chemically-Induced Alterations in Sexual and Functional Development: The Wildlife/Human Connection*, edited by T. Colborn and C. Clement. Princeton, N.J.: Princeton Scientific Publishing.

Holden, A. V., and K. Marsden. 1967. "Organochlorine Pesticides in Seals and Porpoises." *Nature* 216: 1275–76.

Holloway, Marguerite. 1994. "Dioxin Indictment." *Scientific American* 270 (January): 25.

Høyer, A. P., P. Grandjean, T. Jørgensen, J. W. Brock, and H. B. Hartvig. 1998. "Organochlorine Exposure and Risk of Breast Cancer." *Lancet* 352: 1816–20.

Høyer, A. P., T. Jørgensen, J. W. Brock, and P. Grandjean. 2000. "Organochlorine Exposure and Breast Cancer Survival." *Journal of Clinical Epidemiology* 53: 323–30.

Hynes, H. Patricia. 1989. *The Recurring Silent Spring.* New York: Pergamon Press.

Irvine, D. S. 1994. "Falling Sperm Quality." *British Medical Journal* 309: 476.

Jackson, M. B., C. Chilvers, and M. C. Pike. 1986. "Cryptorchidism: An Apparent Substantial Increase since 1960." *British Medical Journal* 293: 1401–4.

Jefferies, D. J. 1975. "The Role of the Thyroid in the Production of Sublethal Effects by Organochlorine Insecticides and Polychlorinated Biphenyls." 131–230 in F. Moriarty, ed. *Organochlorine Insecticides: Persistent Organic Pollutants.* London: Academic Press.

Kolata, Gina. 1988. "PCB Exposure Linked to Birth Defects in Taiwan." *New York Times*, August 2.

Krimsky, Sheldon. 1999. *Hormonal Chaos: The Scientific and Social Origins of the Environmental Endocrine Hypothesis.* Baltimore: Johns Hopkins University Press.

Kurtz, Howard. 2001. "Moyers's Exclusive Report: Chemical Industry Left Out." *Washington Post*, March 22.

"Labeling Cosmetics May Help Prevent Cancers." 2002. Environment News Service. August 15. http://ens-news.com/ens/aug2002/2002-08-15-01.asp.

"*The Lancet* Press Release: Dioxin Exposure Linked to Long-Term Decrease in Male Births." *Lancet*, May 27, 2000.

Laurance, Jeremy. 2000. "Incinerator Pollution Can Have Devastating Effect on Birth Rate." *London Independent*, May 26. (In LEXIS)

Lazaroff, Cat. 2002. "Beauty Products May Contain Controversial Chemicals." July 10. http://ens-news.com/ens/jul2002/2002-07-10-07.asp.

McGinn, Anne Platt. 2000. "POPs Culture." *World Watch*, March–April, 26–36.

Mocarelli, Paolo, Pier Mario Gerthoux, Enrica Ferrari, Donald G. Patterson Jr., Stephanie M. Kieszak, Paolo Brambilla, Nicoletta Vincoli, Stefano Signorini, Pierluigi Tramacere, Vittorio Carreri, Eric J. Sampson, Wayman E. Turner, and Larry L. Needham. 2000. "Paternal Concentrations of Dioxin and Sex Ratio of Offspring." *Lancet* 355 (May 27): 1858–63. http://irptc.unep.ch/pops/newlayout/press_items.htm.

Montague, Peter. 1993. "A New Era in Environmental Toxicology." *Rachel's Environment and Health News* 365, November 25. http://www.rachel.org/bulletin/index.cfm?St = 2.

———. 1994. "PCBs Diminish Penis Size." *Rachel's Environment and Health News* 372, January 13. http://www.rachel.org/bulletin/index.cfm?St = 2.

———. 1997. "The Weybridge Report." *Rachel's Environment and Health News* 547, May 22. http://www.rachel.org/bulletin/index.cfm?St = 2.

Moses, Alan. 2000. "Quality of Sperm Unchanged over 50 Years." Reuters News Service, February 29. Accessed at Web site of the Chlorine Chemistry Council. http://c3.org/news_center/third_party/02-29-00.html.

Moyers, Bill. 2001. "Trade Secrets: A Moyers Report." Program transcript. Public Broadcasting Service, March 26. http://www.pbs.org/tradesecrets/transcript.html.

National Academy of Sciences National Research Council. 1999. *Hormonally Active Agents in the Environment.* Washington, D.C., 2000. http://books.nap.edu/books/0309064198/html/1.html#1.

National Toxicology Program. 2000. *Endocrine Disrupters Low-Dose Peer Review Report.* Research Triangle Park, N.C.: National Institute of Environmental Health Sciences. http://ntp server.niehs.nih.gov/htdocs/liason/lowdosewebpage.html.

Paulozzi, L. J., J. D. Erickson, and R. J. Jackson. 1997. "Hypospadias Trends in Two U.S. Surveillance Systems." *Pediatrics* 100: 831–34.

Peterson, R. E., H. M. Theobald, and G. L. Kimmel. 1993. "Developmental and Reproductive Toxicity of Dioxins and Related Compounds: Cross-Species Comparisons." *Critical Reviews in Toxicology* 23, no. 3: 283–355.

Posten, Lee. 1999. "National Academy of Sciences Report on Hormone Disruptors Released; Growing Evidence of Damaging Health Effects Renews Call for Increased Research into Toxic Chemical Threat." World Wildlife Fund, August 4. http://www.worldwildlife.org/toxics/whatsnew/pr_7.htm.

Reinjders, P. 1986. "Reproductive Failure in Common Seals Feeding off Fish from Polluted Coastal Waters." *Nature* 324: 456–57.

Rogan, Walter J., et al. 1988. "Congenital Poisoning by Polychlorinated Biphenyls and Their Contaminants in Taiwan." *Science* 241 (July 15): 334–36.

"Rural Men Found to Have Poorer Semen Quality." 2002. Associated Press in *Omaha World-Herald*, November 11, 6-A.

Schettler, Ted, Gina Solomon, Maria Valenti, and Anne Huddle. 1999. *Generations at Risk: Reproductive Health and the Environment*. Cambridge, Mass.: MIT Press.

Schwanke, Jane, and Pamela Yoder. 2000. "Malibu Barbie, Holiday Barbie . . . Toxic Barbie? Some Vintage Toys May Ooze Chemical That Could Harm Kids." *WebMD Medical News*, August 25. http://content. health.msn.com/content/article/1728.60731.

Sharpe, R. M. 1993. "Declining Sperm Counts in Men: Is There an Endocrine Cause?" *Journal of Endocrinology* 136: 357–60.

Skakkebaek, N. E. 1972. "Possible Carcinoma-*in-situ* of the Testis." *Lancet* 1(7756) (September): 516–57.

Soto, A. M., K. L. Chung, and C. Sonnenschein. 1994. "The Pesticides Endosulfan, Toxaphene, and Dieldrin Have Oestrogenic Effects on Human Estrogen-Sensitive Cells." *Environmental Health Perspectives* 102: 380–83.

Stevens, J. T., C. B. Breckenridge, L. T. Wetzel, J. H. Gillis, L. G. Luempert, and J. C. Eldridge. 1994. "Hypothesis for Mammary Tumorigenesis in Sprague-Dawley Rats Exposed to Certain Triazine Herbicides." *Journal of Toxicology and Environmental Health* 43: 139–53.

Stewart, Irvine, et al. 1996. "Evidence of Deteriorating Semen Quality in the United Kingdom: Birth Cohort Study in 577 Men in Scotland over 11 Years." *British Medical Journal* 312 (February): 467–71.

Swan, Shanna H., Charlene Brazil, Erma Z. Drobnis, Fan Liu, Robin L. Kruse, Maureen Hatch, J. Bruce Redmon, Christina Wang, James W. Overstreet, and the Study for Future Families Research Group. 2002. "Geographic Differences in Semen Quality of Fertile U.S. Males." Environmental Health Perspectives Online. November 11. http://www. enponline.org/swan2002.

Thomas, Kristin Bryan, and Theo Colborn. "Organo-chlorine Endocrine Disruptors in Human Tissue." *Advances in Modern Environmental Toxicology* 21 (1992): 342–43.

Thornton, Joe. 1997. "PVC: The Poison Plastic." Washington, D.C.: Greenpeace USA, April. http://www.greenpeace.org/~usa/reports/toxics/ PVC/cradle/dcgsum.html.

———. 2000. *Pandora's Poison: Chlorine, Health, and a New Environmental Strategy*. Cambridge, Mass.: MIT Press.

Thrupp, L. A. 1991. "Sterilization of Workers from Pesticide Exposure: The Causes and Consequences of DBCP-Induced Damage in Costa Rica and Beyond." *International Journal of Health Services* 21, no. 4: 731–57.

Toniolo, P. G., M. Levitz, and A. Zeleniuch-Jacquotte. 1995. "A Prospective Study of Endogenous Estrogens and Breast Cancer in Post-menopausal Women." *Journal of the National Cancer Institute* 87: 190–97.

Trichopoulos, Dimitrios. 1990. "Hypothesis: Does Breast Cancer Originate *in Utero? Lancet* 335: 939–40.

Viola, P. L., A. Bigotti, and A. Caputo. 1971. "Oncogenic Response of Rat Skin, Lungs, and Bones to Vinyl Chloride." *Cancer Research* 31: 516–22.

vom Saal, F. S., B. G. Timms, et al. 1997. "Prostate Enlargement in Mice Due to Fetal Exposure to Low Doses of Estradiol or Diethylstilbestrol and Opposite Effects at High Doses." *Proceedings of the National Academy of Sciences* 94: 2056–61.

Wetzel, L. T., L. G. Luempert III, M. O. Breckenridge, J. T. Stevens, A. K. Thakur, P. J. Extrom, and J. C. Eldridge. 1994. "Chronic Effects of Atrazine on Estrus and Mammary Tumor Formation in Female Sprague-Dawley and Fischer 344 Rats." *Journal of Toxicology and Environmental Health* 43: 169–82.

Whorton, M. D., and D. E. Foliart. 1983. "Mutagenicity, Carcinogenicity, and Reproductive Effects of Dichloropropane (DBCP)." *Mutation Research* 123: 13–30.

Wolfe, W. H., J. E. Michalek, and J. C. Miner. 1994. "Determinants of TCDD Half-life in Veterans of Operation Ranch Hand." *Journal of Toxicology and Environmental Health* 41: 481–88.

World Wildlife Fund. 2000. "Toxics—What's New." http://www.worldwild life.org/toxics/whatsnew/pr_7.htm.

# 8

# Solutions: Public Policy Issues

## THE STOCKHOLM CONVENTION ON PERSISTENT ORGANIC POLLUTANTS: AN INTERNATIONAL BAN ON POPS

While world diplomacy has thus far proved inadequate to the challenges of global warming, POPs have been rather quickly banned by an international agreement that is being ratified as this book is being written. Evidence of health damage from POPs (especially among the Inuit and animals of the Arctic) is so easily recognizable that even the George W. Bush White House quickly announced support for the Stockholm Convention to eliminate POPs. Such support does not begin the endgame for the Stockholm Convention, however. During the next several years, meeting the goals of the convention will require basic changes in the manufacturing processes of many products that are used by many millions of people worldwide. The Stockholm Convention eventually will require entire industries to find new ways to produce even basic commodities such as steel and to dispose of waste without emitting the newly banned by-products.

The Stockholm Convention requires that nine of the dirty dozen be removed completely from use and production: aldrin, chlordane, toxaphene, dieldrin, endrin, heptachlor, hexachlorobenzene, mirex, and toxaphene. Exceptions to complete elimination have been allowed for dioxins and furans, which are to be reduced immediately and eventually eliminated "where feasible," by restrictions on open trash burning, for example (Kaiser and Enserink 2000, 2053).

The Stockholm Convention provides an exemption for DDT when it is used to control malaria. Malaria killed 2.7 million people worldwide in 1999; it is the world's single deadliest disease, claiming more lives than AIDS/HIV, and its toll is growing. Malaria experts testified at the Stockholm negotiations that no feasible substitute exists for malaria control, DDT's original use during World War II. About twenty-five countries still find DDT necessary for use against malaria. The Stockholm Convention urges intensified research into identifying alternatives to DDT. Countries using DDT are required to document who is utilizing it, and under what circumstances, in a special register.

Even with a world protocol that aims to eliminate them, the dirty dozen (and many other organochlorines) will be with us as serious environmental pollutants for several decades to come at least. Not only are they long-lasting in the atmosphere, but industries that produce them have become expert at semi-green shuck and jive that announces compliance while ducking speedy implementation of new rules.

Representatives from about one hundred twenty countries agreed on the Stockholm Convention during December 2000. Five months later, on the eve of Earth Day 2001, U.S. President George W. Bush agreed to sign the Stockholm Convention on Persistent Organic Pollutants and said he would submit it to the U.S. Senate for ratification. Coming within a month of Bush's repudiation of the Kyoto Protocol (meant to reduce emissions of greenhouse gases), the Stockholm Convention has some diplomatic virtues from the standpoint of U.S. interests. First, the United States, which already had banned many of the dirty dozen, has little to lose, in contrast with any meaningful effort to reduce global warming. The production of carbon dioxide and methane (the two main greenhouse gases) are intricately wedded to our fossil-fueled industrial base.

The Stockholm Convention was the result of nearly two decades of international diplomacy aimed at restricting and eventually ending the manufacture and use of the most notorious POPs. With their threat to stratospheric ozone becoming more evident during the 1980s, CFCs were the first major organochlorine to be restricted by the Montreal Protocol in 1987.

At the 1992 Earth Summit in Rio de Janeiro, more than one hundred seventy governments committed in their Agenda 21 to eliminate emissions of organohalogen and other synthetic compounds that threatened to accumulate to dangerous levels. The UN Environment Program's (UNEP) May 1995 Governing Council agreed to initiate an expedited assessment of the twelve priority POPs and their alternatives. In June 1995 the governments of Canada and the Philippines held an international experts' meeting on POPs in Vancouver. The final consensus statement of that meeting stated that "There is enough scientific information on the adverse human health and environmental impacts of POPs to warrant coherent action at the national, regional, and international level. This will include bans, phase-outs and provisional severe restrictions for certain POPs" (World Wildlife Fund 2000).

With this scientific consensus in hand, a global UNEP conference convened in November 1995 in Washington, D.C. Although its focus was on protection of the marine environment from land-based activities, special attention was devoted to POPs, with a high-level ministerial segment agreeing by consensus that "International action is needed to develop a global, legally binding instrument, amongst other international and regional actions, for the reduction and/or elimination of emissions and discharges, whether intentional or not, and where appropriate, the elimination of the manufacture and use of [the twelve priority POPs]" (World Wildlife Fund 2000).

The Intergovernmental Forum on Chemical Safety (IFCS) developed recommendations (Buccini 1996) in 1996 that concluded that sufficient evidence existed to warrant a global treaty to minimize the risk from the twelve specified POPs. IFCS called for immediate action by UNEP and the World Health Assembly to reduce or eliminate POPs emissions and discharges. In February 1997 the UNEP Governing Council endorsed IFCS's recommendations and agreed by consensus to move forward with treaty negotiations.

During these early negotiations on an international ban of POPs, officials from the World Health Organization (WHO) and delegates from several developing countries questioned the elimination of DDT because of its major role in combating malaria and other insect-borne diseases. Malaria poses a threat to at least 2.5 billion people in more than ninety countries and contributes every year to about three million deaths, over half of them children under five years old. Although the WHO and its experts have slowly embraced disease-fighting methods that reduce the reliance on DDT, African delegates stress the need to find and fund cost-effective alternatives.

Negotiators also faced the question of how to identify, collect, and destroy POPs that remain in obsolete stockpiles of persistent chemicals or in hot spots of environmental contamination. In a number of developing countries, obsolete pesticides, including POPs, are stored in extremely hazardous conditions, as are old PCB-containing transformers and capacitors.

The Stockholm Convention seeks, within some practical limitations, to realize a point of view enunciated by the World Wildlife Fund:

Our decades of experience with persistent chemicals have demonstrated unequivocally that there is no way to manage POPs. The only responsible course is to eliminate their production, use, and release as quickly as possible, while recognizing and addressing the special circumstances of developing countries in need of assistance. The time has come to stop this experiment with "hand-me-down poisons" before it does more irreparable damage to wildlife, children, and adults. (World Wildlife Fund 2000)

The Stockholm Convention on Persistent Organic Pollutants will be legally binding once 50 of the 122 signatory nations have ratified it, a process that could take several years. The new protocol also contains mechanisms for adding other chemicals to be banned in future years.

The Stockholm Convention requires rich nations to pay developing countries to phase out the banned pollutants and to destroy stockpiles, mostly within five years of ratification. Use of these pollutants after that time will require stringent proof of need. In the case of PCBs, which have been widely used in electrical transformers and other equipment, governments may maintain existing equipment in a way that prevents leaks until 2025 to allow time to arrange for PCB-free replacements. Although PCBs are no longer produced, hundreds of thousands of tons are still used in such equipment. In addition, a number of country-specific and time-limited exemptions have been established for other chemicals. The Stockholm Convention also includes plans for a $150 million fund to be managed by the United Nations that will help Third World countries dispose of toxic stockpiles of gift waste (see chapter 4).

Experts opened talks in mid-January 2002 in Geneva to develop policies and technical guidelines for a major treaty on transporting and discarding toxic pollutants under the Stockholm Convention banning POPs and other programs under the aegis of the United Nations. The convention's Technical Working Group developed technical guidelines for environmentally sound management of waste lead-acid batteries, metal and metal compounds, plastic wastes, and POPs and the dismantling of ships. These issues are covered by the Basel Convention on the Control of the Transboundary Movements of Hazardous Wastes and Their Disposal.

The Basel Convention was adopted in March 1989 after a series of notorious toxic cargoes from industrialized countries drew public attention to the dumping of hazardous wastes in developing and East European countries. The treaty, which has 149 parties, obliges its members to ensure that such wastes are managed and disposed of in an environmentally sound manner. Governments are expected to minimize the quantities that are transported, to treat and dispose of wastes as close as possible to where they were generated, and to minimize the generation of hazardous waste ("Hazardous Waste Treaty" 2002).

The day before these discussions began, Turkish police arrested seventeen Greenpeace activists who had occupied a Swiss ship at a ship-breaking yard in Aliaga, Turkey. Unfurling a banner that said "Stop Toxic Shipbreaking" from on board the *Star of Venice,* the Greenpeace representatives demanded an end to the practice of scrapping ships that contain toxic materials on Turkish beaches.

Even as international diplomacy works out an eventual ban of them, release of POPs into the environment is continuing, and a possibility exists for further severe impacts on the health of wildlife and humans. Of particular concern are effects on the developing stages of life, the unborn and nursing young. Furthermore, in addition to POPs targeted by international diplomacy, there are many other POPs and hazardous chemicals that are toxic to health. Many POPs remain in the atmosphere (notably in the Arctic and Antarctic) for a century or more, and so the struggle to restore the ecosystem will be long and difficult.

In mid-April 2002 the Bush administration submitted the Stockholm Convention on Persistent Organic Pollutants to the U.S. Senate for ratification. Additional legislation to amend existing U.S. laws needed to implement the POPs treaty and two related treaties also was directed to Congress at the same time. The Bush legislation restricts the treaty in one important way, however: it lacks provisions contained in the Stockholm Convention to add other chemicals to the banned list in the future. "The Bush administration proposal ties the Environmental Protection Agency's hands, limiting domestic implementation to 12 POPs already regulated in the U.S.," said public health expert Robert Musil, who is chief executive officer and executive director of Physicians for Social Responsibility. "Congress must now come forward with legislation that will implement the full intent of the POPs treaty as a dynamic, flexible instrument to protect human health now and into the future," he urged ("POPs Treaty" 2002).

## PROBLEMS WITH "PROOF"

The Stockholm Convention was negotiated despite arguments on some fronts that participants lacked absolute scientifically valid proof that the dirty dozen are environmentally harmful. The Weybridge Report outlines some of the problems with such a standard of proof:

It is not possible fully to understand the significance of levels found in tissues or blood until the mechanism and timing of action of EDS [endocrine disrupting substances] and their metabolites are better understood. . . . It is not anticipated that any useful SAR [structure-activity relationships] . . . will emerge." (*European Workshop* 1997, 1: 37, 45)

The same report also stresses (1: 41) that "potential interactive effects of exposure to several substances simultaneously need to be taken into account."

The structures of endocrine-disrupting chemicals vary widely, and so, according to Peter Montague, "No structure-activity relationships are likely to emerge to help scientists decide which ones are bad actors. This means all chemicals are candidates for testing, which greatly complicates (and boosts the price of) testing to find endocrine disrupters" (Montague 1997). Test-tube examination of chemicals will not yield the needed information, according to Montague, who has written that tests must be done on living animals that are expensive and often cruel (Montague 1997).

The necessary tests for the thousand most common toxic chemicals in unique combinations of three "would require at least 166 million different experiments," according to Montague (1997). This means that endocrine disrupters could have substantial effects that will demonstrate acceptable proof under laboratory standards. Montague described the absurdity of the situation:

If we wanted to conduct the 166 million experiments in just 20 years, we would have to complete 8.3 million tests each year. The U.S. presently has the capacity to conduct only a few hundred such tests each year. Just training sufficient personnel to conduct 8.3 million animal tests each year is beyond our national capacity. (Montague 1997)

Because verifiable scientific proof is so elusive, the Weybridge Report recommends use of the Precautionary Principle (1: 52). Developed at the Rio Earth Summit (1992), the Precautionary Principle holds that "Where there are threats of serious or irreversible damage, lack of full scientific certainty shall not be used as a reason for postponing cost-effective measures to prevent environmental degradation" (Bodansky 1994, 203). The chemical industry has long argued that harm should be proven before products are removed, but environmentalists believe the search for a largely unachievable degree of scientific certainty is a corporate stalling tactic.

## REQUIRE COMPLICIT COMPANIES TO PAY FOR CLEANUP

During 2002, companies found responsible for PCB contamination in Anniston, Alabama, were required to pay costs associated with its cleanup. In late March the U.S. Justice Department and the U.S. EPA reached an agreement with Solutia, Incorporated (formerly Monsanto Company) and Pharmacia Corporation to clean up PCBs that leaked from a former Monsanto plant in western Anniston. The PCBs continued to contaminate air and water of several low-income neighborhoods, regional waterways, and about forty miles of floodplain in the area. The corporate-funded cleanup was in lieu of a Superfund listing for the community, requiring tax-assisted cleanup. According to an account by the Environment News Service, Solutia and Pharmacia agreed to emergency cleanup of the worst residential contamination in the area.

Under the pact, according to ENS, Solutia and Pharmacia were required hire EPA-approved contractors to study all contaminated areas for PCBs and other environmental pollutants, and evaluate the risk they pose to public health and the environment. The study will determine available cleanup options and suggest a strategy for restoring the community, covering all areas where PCBs have been found, including the Solutia facility, the landfills, creeks, rivers, lakes, flood plains and residential, commercial and agricultural properties that surround the facility ("Settlement" 2002). The agreement also includes establishment of a $3.2 million foundation for special-education needs of children in the Anniston area. The companies also were required to pay the EPA as much as $6 million for the agency's investigative costs.

Some community activists (such as members of the Anniston-based Community against Pollution) feared, however, that the complicit companies may be using the agreement as a stalling tactic. Their fears were supported when Solutia filed a petition in court asking that a lawsuit seeking a court-ordered cleanup be dismissed.

## SEEKING A BAN OF PVCS

At the top of many environmental activists' lists of POP-related business is a ban of PVC production. Greenpeace recommends a phase-out of PVC production and use, and the European Union has taken a major step toward that goal (see chapter 7). As Joe Thornton has written:

The health risk posed by dioxins calls for immediate action to reduce and ultimately eliminate the production and use of PVC. PVC is the single largest use of elemental chlorine and its production is expanding. It is also known that dioxin is generated as a byproduct during PVC production, use, or treatment for disposal. There are strong grounds for holding PVC responsible for a substantial and growing proportion of global dioxin production and release. . . . Reducing the production and use of PVC is a simple and effective avenue to prevent PVC-related dioxin pollution. By replacing PVC with alternative, chlorine-free materials, dioxin formation associated with PVC can be eliminated entirely. Given the importance of the PVC lifecycle in the nation's dioxin burden, a PVC phase-out must be a top priority in any dioxin-prevention strategy. (Thornton 1997)

According to another observer, "The single most effective way to lower dioxin output is to get as much chlorine as possible out of the waste stream. PVC is the source of an estimated 80 percent of the chlorine that flows into municipal waste incinerators and nearly all the chlorine in medical waste incinerators" (McGinn 2000, 36).

Numerous businesses have eliminated or begun working toward a PVC phase-out in their products and facilities, including Nike, Volvo, Saab, Braun, Ikea, the Body Shop, and Svenska Bostder (one of Sweden's leading construction companies). Major construction projects such as the Sydney 2000 Olympic village were designed to minimize the use of PVCs "by selecting alternative materials where they are available, are fit for the purpose and are cost-competitive" (David Richmond, director general of the Sydney 2000 Olympic Coordination Authority, to Nicole Williams, chief executive of the Plastics and Chemicals Industries Association, May 15, 1998, cited in Cray and Hardin 1998).

## CHANGE THE EPA'S REVIEW PROCEDURES

Scientists and other experts who advise the EPA on regulations governing toxic chemicals sometimes maintain ties with the affected industries (among other conflicts of interest), according to a study by the General Accounting Office (GAO). Congress established EPA science advisory boards in 1978 to provide independent analysis via peer review in a balanced manner. The report asserted that the EPA had failed to probe adequately for panel members' ties to the chemical industry.

According to the *Washington Post,* "The report found serious deficiencies in the EPA's procedures for preventing conflicts of interest and ensuring a proper balance of views among members of Science Advisory Board panels. For example, four of the 13 panel members who studied the cancer risks of the toxic chemical 1,3-butadiene in 1998 had worked for chemical companies or industry-affiliated research organizations—including one who had worked for a company that manufactured 1,3-butadiene, according to the report" (Pianin 2001, A-2).

The GAO identified similar problems with three other cancer-risk-assessment panels. "In one case," according to the *Post* report, "Seven of 17 advisory-board members worked for chemical companies or for industry-affiliated research organizations. Five other panelists had received consulting or other fees from chemical manufacturers" (Pianin 2001, A2). "The regulatory process benefits from scientific and technical knowledge, expertise and competencies of panel members," the report stated. "However, the work of fully competent peer review panels can be undermined by allegations of conflict of interest and bias" (Pianin 2001, A2). "The American people expect decisions that affect environmental and public health regulations to be based on unbiased science," Rep. Henry A. Waxman, Democrat of California, said, "but this GAO study reveals polluting industries are in a position to influence panel findings" (Pianin 2001, A2).

Before the release of this report, Greenpeace and the Center for Health, Environment, and Justice already had complained about

the composition of an EPA panel convened to study the health effects of dioxin. About one-third of the twenty-one panel members
had worked as paid consultants to the chemical industry, according
to the GAO study (Pianin 2001, A2). Erik Olson of the Natural Resources Defense Council, an environmental group, said that "we're
seeing . . . advisory board panels—stacked with industry mouthpieces—acting like kangaroo courts to strike down important EPA
initiatives" (Pianin 2001, A2). The panel's composition was a major
reason that the EPA did not officially recognize the carcinogenic nature of dioxins until 2000, long after the U.S. armed forces began
giving medals to veterans who suffered cancers (and other maladies)
from Agent Orange toxicity three decades earlier.

As controversy surfaced at EPA regarding possible conflicts of interest on its review panels, questions were being raised during the
summer of 2001 about President George W. Bush's nominee to head
an obscure but powerful White House agency, the Office of Information and Regulatory Affairs, which is part of the Office of Management and the Budget. Its administrator acts as a regulatory czar
who can block any new regulation, whether arsenic standards for
drinking water, worker-protection rules, or the public's right to
know (Durbin 2001).

As a participant in the aforementioned EPA Science Advisory
Board subcommittee on dioxin, John Graham, Bush's nominee,
had said that reducing dioxin levels too far might "do more harm
. . . than good." Graham also asserted that dioxin might prevent
cancer in some cases, a hypothesis that a senior EPA official called
"irresponsible and inaccurate." Graham also told Congress that
smog protects people from the harmful effects of too much sunlight.
Graham also rejected the idea that pesticides in foods may pose
health problems, calling this a "trivial" concern that reflects public
"paranoia" about toxic chemicals (Durbin 2001, A15).

"There is no justification for completely abandoning chemicals *per
se* as components in the defensive toolbox for managing pests"
(Committee on the Future Role 2000, 253). "A concerted effort in
research and policy should be made to increase the competitiveness
of alternatives to chemical pesticides" (Committee on the Future
Role 2000, 255). The need for alternative methods is underscored
by the fact that "pesticide resistance now is universal across taxa"
(Committee on the Future Role 2000, 257).

## TAXATION

A tax has been proposed on the chlor-alkali process per ton of chlorine produced. Revenue, under this proposal, would be held in a fund to aid the transition to a chlorine-free industrial society. In particular, funds should be used for exploring and demonstrating economically viable alternatives and for easing dislocations among affected workers and communities. In addition, funds should be targeted toward clean production processes to assist developing countries and small businesses in making the change (Allsopp et al. 1995).

## EATING ORGANICALLY

The organic-food market in the United States generated $178 million in 1980, $2.8 billion in 1995, and more than $5.4 billion in 1998. There were 2,841 U.S. organic farms in 1991, 4,060 in 1994, and 12,000 in 1997 (Committee on the Future Role 2000, 113). By the late 1990s, organic-food sales were increasing 20 percent a year in the United States (Committee on the Future Role 2000, 145). The sustained increase in of the market for organic produce is undoubtedly a reaction, at least in part, to rising anxiety about organochlorines in food.

## GOING NONTOXIC: WISCONSIN'S POTATO FARMERS

Since 1995 Wisconsin potato farmers have made a concerted effort to slash their usage of several toxic chemicals, reducing their consumption of them by a half million pounds in three years. The reductions were described in a report by the World Wildlife Fund, the Wisconsin Potato and Vegetable Growers Association, and the Integrated Pest Management team at the University of Wisconsin ("A Report" 2000). Wisconsin potato farmers have reduced their use of eleven high-risk pesticides.

The farmers used integrated pest-management techniques such as scouting and spot-treating for pests, computer prediction models to determine when pests are most active, and intensive crop rotations. Some of the farmers also used newer, less toxic pesticides. Wisconsin farmers annually plant 80,000 acres of potatoes, making

Wisconsin the third-largest potato-producing state ("A Report" 2000).

The Wisconsin farmers acted early to reduce vulnerability to regulation, address the increasing problem of pesticide resistance, and protect the quality of wildlife habitat. The Wisconsin program could provide a model for other farmers in other states and nations.

## PULP AND PAPER: GETTING THE CHLORINE OUT

Chlorine use by paper producers has been declining significantly. The paper-making industry once consumed one-sixth of all chlorine produced, making it the second-largest chlorine-using industry (PVC plastic manufacture is the largest). Pulp and paper mills that use older chlorine-gas bleaching methods create and release highly toxic and persistent organochlorine chemicals.

Elimination of chlorine-based bleaching by the paper industry is well underway in Europe. Most of the twenty-eight mills worldwide that were producing commercial quantities of chlorine-free kraft pulp by 1994 were located in Europe. In 1993, 34 percent of all printing paper (excluding newsprint) sold in the Germanic and Scandinavian countries of Europe was chlorine-free. Industry sources forecast that by 1996, this percentage will double to 68 percent (Palter and Weinberg 1994). The German weekly newsmagazine *Spiegel* converted to chlorine-free paper in response to growing pressure from environmentalists. Other magazine publishers followed.

Some book publishers have been purchasing only chlorine-free paper. The Cornell University Press says as much on the copyright pages of its books:

Cornell University Press strives to use environmentally responsible suppliers and materials to the fullest extent possible in the publishing of its books. Such materials include vegetable-based, low-VOC inks and acid-free papers that are recycled, totally chlorine-free, or partially composed of non-wood fibers. Books that bear the logo of the FSC (Forest Stewardship Council) use papers taken from forests that have been inspected and certified as meeting the highest standards for environmental and social responsibility. (Huhndorf 2001, copyright page)

For several years an American Library Association task force on the environment tried to get the ALA Council to endorse a resolution advocating use of chlorine-free papers, which may be bleached with oxygen or hydrogen peroxide. After almost four years, this standard was largely adopted.

As a result of bleaching process changes and strict EPA standards, releases of dioxins from the nation's nearly one hundred bleached-pulp mills have dropped by at least 94 percent since 1988. Dioxins have been eliminated from the wastewaters of some mills that have switched to largely chlorine-free technology.

## BANNING BACKYARD TRASH INCINERATION

One substantial remaining uncontrolled source of dioxin release into the environment is residential waste incineration, including backyard burning of trash, most notably in rural areas without garbage pickup service. Such burning is "one of the largest unaddressed dioxin sources and one that could have a disproportionately large contribution to the [toxicity of the] food supply," the U.S. EPA has stated (Skrzycki and Warrick 2000, A1).

## A PCB-EATING BACTERIUM?

Researchers have discovered a strain of bacteria capable of breaking down the toxic PCBs that contaminate soil and sediments near many industrial sites. The bacterium breaks down tough chlorine bonds in PCBs in river and harbor sediments. The discovery of the bacterium was reported in *Environmental Microbiology* by scientists with the University of Maryland Biotechnology Institute (UMBI) ("Bacteria Breaks Down" 2002).

In experiments using bottom sediments from Baltimore Harbor, researchers of UMBI's Center of Marine Biotechnology (COMB) and the Medical University of South Carolina (MUSC) discovered the PCB-degrading bacterium using a rapid DNA screening method. According to a report by the Environment News Service, "Particles of PCBs persist after many years, because they don't dissolve well in water. They attach to sediment and get covered over," said Kevin Sowers, research microbiologist at COMB. "Unless there is some turnover, a lot of PCBs stay hidden" ("Bacteria Breaks Down" 2002).

"This first identification of a PCB dechlorinating, anaerobic [without oxygen] bacterium is important for bioremediation efforts and for developing molecular probes to monitor PCB degrading where they are found," said Sowers ("Bacteria Breaks Down" 2002). The researchers linked PCB dechlorination to the growth of the bacterium, which appears to live off the compound. "This is a great example of how manmade pollution can be handled by microorganisms through their incredible ability to adapt," said Jennie Hunter-Cevera, UMBI president and environmental biotechnologist ("Bacteria Breaks Down" 2002). The report concludes that the UMBI method could be used to identify additional PCB-degrading microbes.

## PROSECUTING THE FRIO BANDITOS

By 2002 CFC smuggling was becoming more than a legal curiosity, with more cases coming to light. By the early years of the third millennium, according to the U.S. Environmental Investigation Agency, the Frio Banditos were serving a multimillion-dollar international market that spans the United States and several Third World nations. Transactions had reached 20,000 tons a year ("Smuggled C.F.C.s" 2002).

A Florida man faced a fine and possible jail time for smuggling CFCs into the United States. Clifford Windsor of Fort Lauderdale, pleaded guilty on December 19, 2001, to violating the Clean Air Act by illegally importing about three hundred cylinders of Freon-12, a CFC refrigerant, some of which were sold to businesses. The conviction carries a maximum sentence of up to five years in prison and a fine of up to $250,000. The case was investigated by the U.S. EPA's Criminal Investigation Division, the U.S. Coast Guard, the U.S. Customs Service, and the Ft. Lauderdale Police Department ("Smuggled CFCs" 2002). Before the Montreal Protocol, the refrigerant R-12 was used legally in air conditioners in automobiles manufactured before 1994.

## GARDENING WITHOUT CHEMICAL PESTICIDES AND HERBICIDES IN CANADA

Early in 2002, Canada's largest food distributor made a public commitment to stop marketing chemical pesticides within a year, by the beginning of the growing season in 2003. Loblaw Companies Limited announced that it would no longer sell chemical pesticides in its 440 garden centers. "In response to overwhelming consumer demand to eliminate the cosmetic use of pesticides in home gardens, Loblaw Companies Limited has decided to discontinue the sale of chemical pesticides in our garden centers, starting with the spring season 2003," said Loblaw spokesperson Geoff Wilson ("Giant Canadian" 2002).

The company timed its announcement to coincide with the March 14, 2002, opening of the annual Canada Blooms Flower and Garden Show at the Metro Toronto Convention Centre, of which Loblaw is the primary sponsor. "We will use this venue to launch our commitment to Chemical Pesticide Free by 2003 and to further educate gardeners on how they can enjoy a chemical pesticide-free garden," Wilson said ("Giant Canadian" 2002). According to Environment Canada, a federal agency, homeowners use only 5 to 10 percent of all pesticides sold in Canada, but they tend to use them inappropriately more often than commercial or agricultural users. The Environment News Service reported that the agency advised Canadian homeowners to keep the use of hazardous chemicals to a minimum.

The Sierra Club of Canada joined Loblaw in advising that many pesticides cause damage to wild species. The Sierra Club said that Carbofuran, a Canadian brand, was banned for most uses during the late 1990s, after it was found to be responsible for the near-extinction of the burrowing owl. "Evidence indicates that even the family dog is a victim of pesticides. Dogs from homes with lawns that have been sprayed with pesticides have a higher than average rate of the canine equivalent of lymphoma. Cancer is now the number one cause of death in dogs," the Sierra Club said ("Giant Canadian" 2002).

Loblaw Companies' garden centers have carried some organic-based gardening products in the past. Last year, the company approached its suppliers to develop more organic-based gardening products to replace their chemical pesticides. "By the spring of 2003, we will have organic alternatives for virtually all of the chemical pesticides that we currently carry," Wilson said. Company staff will provide educational handouts and information to consumers on how they can reduce their dependency on chemical pesticides ("Giant Canadian" 2002).

During 2001, according to the Environment News Service, the Supreme Court of Canada established a legal right for municipalities to forbid the use of chemical pesticides on public and private property within their jurisdictions. On June 28, 2001, the Supreme

Court upheld Bylaw 207 of the town of Hudson, Quebec, which bans pesticide use on public and private property for aesthetic purposes. The bylaw had been challenged in the Quebec courts and then in the Canadian Supreme Court by lawn-pesticide companies after they were charged with violating the ban ("Giant Canadian" 2002).

## PHYTOREMEDIATION: PLANTS THAT ABSORB HEAVY METALS AND ORGANIC CHEMICALS

Also during 2002, almost $2.22 million worth of grants was awarded to seven universities to study the ability of plants to treat soils contaminated by heavy metals or organic chemicals. The research studied phytoremediation, which was described by the Environment News Service as "The use of plants to degrade, remove or stabilize toxic compounds from contaminated soil and water in ways that are less expensive and less disruptive than traditional cleanup techniques" ("$2.2 Million" 2002).

The U.S. EPA and the National Science Foundation sponsored the grants to accelerate the development of innovative scientific solutions to the worldwide problem of soil contamination by heavy metals and organic chemicals. Funding for the joint initiative was made available through the Joint Program on Phytoremediation, a U.S. federal government research effort involving the EPA, the National Science Foundation, and the Departments of Defense and Energy.

Three grants were awarded through EPA's Science to Achieve Results (STAR) program, designed to clarify the mechanism of phytoremediation of organic contaminants. These three grants were given to the University of California, University of Connecticut, and Washington State University. The University of California at Riverside received money to evaluate plant species that produce a specific group of chemicals for use in phytoremediation and to study the ecology of chemical-degrading bacteria that live in the root systems of these plants. The University of Connecticut was funded to investigate the role of plant roots in the phytoremediation of POPs in soil ("$2.2 Million" 2002). Washington State University will study the potential use of spartina cordgrasses as a phytoremediation tool in marine and estuarine sediments.

The National Science Foundation sponsored three multidisciplinary research projects to investigate the genetic components of phytoremediation of heavy metals in soils. A grant went to Cornell

University, one went to Purdue University, and a joint grant went to Northwestern University and the University of Florida. Cornell was funded to study the molecular basis for heavy-metal accumulation and tolerance in one plant species. Purdue University attempted to identify genes for metal accumulation in an entire plant genome, the Brassicaceae family ("$2.2 Million" 2002). Northwestern University and the University of Florida performed research to clarify the mechanisms of arsenic uptake, translocation, distribution, and detoxification by brake ferns.

More information on these grants is available at http://www.nsf.gov/bio/ibn/ibndevelop.htm and http://es.epa.gov/ncer/grants/phyto01.html.


## A TRUE GREEN OXIDANT

Scientists and the chemical industry have been testing new catalytic oxidation systems that use metal complexes to break down some POPs before they can pollute the atmosphere. Sayam Sen Gupta and colleagues have reported in *Science* that a new oxidative cleaning method may be useful in reducing pollution caused by chlorinated phenols. The system uses hydrogen peroxide and iron catalysts to eliminates between 90 and 98 percent of the offending chemicals, without producing any other toxic by-products (Sen Gupta et al. 2002; Meunier 2002). Much of the resulting mixture (about 83 percent) is released as benign free chlorine. An analyst of the process finds that it is much more efficient than biodegradation, commenting that "An efficient catalytic chemical method for the oxidative degradation of chlorinated phenols, such as reported by Sen Gupta, et al., is therefore highly welcome for treating industrial wastewater before it is released" (Meunier 2002, 271). The nontoxic nature of this process, according to Meunier, makes it a true green oxidant.


## REFERENCES

Allsopp, Michelle, Ben Erry, Ruth Stringer, Paul Johnston, and David Santillo. 2000. "Recipe for Disaster: A Review of Persistent Organic Pollutants in Food." Exeter, U.K.: Greenpeace Research Laboratories and University of Exeter Department of Biology. http://www.green peace.org/~toxics/reports/recipe.html.

"Bacteria Breaks Down PCBs in Baltimore Harbor." 2002. Environment News Service, January 8. http://ens-news.com/ens/jan2002/2002L-01-08-09.html.

Bodansky, Daniel. 1994. "The Precautionary Principle in U.S. Environmental Law." In *Interpreting the Precautionary Principle,* edited by Timothy O'Riordan and James Cameron. London: Earthscan Publications.

Buccini, John. 1996. "Recent Developments in the Intergovernmental Forum on Chemical Safety (IFCS)," citing *Final Report of the Intergovernmental Forum on Chemical Safety ad hoc Working Group on Persistent Organic Pollutants.* United Nations Environmental Programme, POPs Team. Geneva, Switzerland. July 1. http://www.chem.unep.ch/pops/POPs_Inc/proceedings/stpetbrg/buccini.htm.

Committee on the Future Role of Pesticides in U.S. Agriculture, Board on Agriculture and Natural Resources, Board on Environmental Studies and Toxicology, Commission on Life Sciences, and National Research Council. 2000. *The Future Role of Pesticides in U.S. Agriculture.* Washington, D.C.: National Academies Press.

Cray, Charlie, and Monique Harden. 1998. "PVC and Dioxin: Enough Is Enough." *Rachel's Environment and Health News* 616, September 18. http://csf.colorado.edu/envtecsoc/98/0285.html.

Durbin, Dick. 2001. "Graham Flunks the Cost-Benefit Test." *Washington Post,* July 16. http://www.washingtonpost.com/wp-dyn/articles/A1421-2001Jul15.html.

European Commission. 1997. *European Workshop on the Impact of Endocrine Disruptors on Human Health and Wildlife, 2–4 December, 1996, Weybridge, U.K.: Report of Proceedings.* Copenhagen, Denmark: European Commission.

"Giant Canadian Food Chain Rejects Chemical Pesticides." 2002. Environment News Service, March 12. http://ens-news.com/ens/mar2002/2002L-03–12–01.html.

"Hazardous Waste Treaty Gets a Makeover." 2002. Environment News Service, January 14. http://www.ens-news.com/ens/jan2002/2002L-01-14-02.html.

Huhndorf, Shari M. 2001. *Going Native: Indians in the American Cultural Imagination.* Ithaca, N.Y.: Cornell University Press.

Kaiser, Jocelyn, and Martin Enserink. 2000. "Treaty Takes a POP at the Dirty Dozen." *Science* 290 (December 15): 2053.

McGinn, Anne Platt. 2000. "POPs Culture." *World Watch,* March–April, 26–36.

Meunier, Bernard. 2002. "Catalytic Degradation of Chlorinated Phenols." *Science* 296 (April 12): 270–71.

Montague, Peter. 1997. "The Weybridge Report." *Rachel's Environment and Health News* 547, May 22. http://www.rachel.org/bulletin/index.cfm?St=2.

Palter, Jay, and Jack Weinberg. 1994. *Just a Matter of Time: Corporate Pollution and the Great Lakes.* Greenpeace International Chlorine-Free Campaign, Washington, D.C., July. http://www.greenpeace.org/~toxics/reports/reports.html.

Pianin, Eric. 2001. "Toxic Chemical Review Process Faulted; Scientists on EPA Advisory Panels Often Have Conflicts of Interest, G.A.O. Says." *Washington Post,* July 16. http://www.washingtonpost.com/wp-dyn/articles/A59494-2001Jul13.html.

"POPs Treaty Goes to U.S. Senate for Ratification." 2002. Environment News Service, April 11. http://ens-news.com/ens/apr2002/2002L-04-11-02.html.

"A Report by the World Wildlife Fund, the Wisconsin Potato and Vegetable Growers Association, and the Integrated Pest Management Team, at the University of Wisconsin." 2000. July 18. http://www.worldwildlife.org/toxics/whatsnew/pr_17.htm.

Sen Gupta, Sayam, Matthew Stadler, Christopher A. Noser, Anindya Ghosh, Bradley Steinhoff, Dieter Lenior, Colin P. Horwitz, Karl-Werner Schramm, and Terrence J. Collins. 2002. "Rapid Total Destruction of Chlorophenols by Activated Hydrogen Peroxide." *Science* 296 (April 12): 326–28.

"Settlement Will Continue PCB Cleanup in Anniston." 2002. Environment News Service, March 25. http://ens-news.com/ens/mar2002/2002L-03–25–09.html.

Skrzycki, Cindy, and Joby Warrick. 2000. "EPA Links Dioxin to Cancer: Risk Estimate Raised Tenfold." *Washington Post,* May 17. http://irptc.unep.ch/pops/newlayout/press_items.htm.

"Smuggled CFCs Convict Florida Man." 2002. Environment News Service. January 11. http://ens-news.com/ens/jan2002/2002L-01-11-09.html.

Thornton, Joe. 1997. "PVC: the Poison Plastic." Washington, D.C.: Greenpeace USA, April. http://www.greenpeace.org/~usa/reports/toxics/PVC/cradle/dcgsum.html.

"$2.2 million Supports Phytoremediation Studies." 2002. Environment News Service, March 12. http://ens-news.com/ens/mar2002/2002L-03-12-09.html.

U.S. General Accounting Office. 2001. "Environmental Protection Agency's Science Advisory Board Panels: Improved Policies and Procedures Needed to Ensure Independence and Balance." http://www.gao.gov/new.items/d01536.pdf.

World Wildlife Fund. 2000. "Toxics—What's New." http://www.worldwildlife.org/toxics/whatsnew/pr_7.htm.

# Glossary

*Agent Orange:* A dioxin-based defoliant (herbicide) widely used by the United States during the Vietnam War.

*Aldrin:* An organochlorine pesticide; one of the dirty dozen to be banned under the Stockholm Convention.

*Anthropogenic:* Caused or influenced by human activity.

*Atrazine:* Trade name of the most-often-used herbicide in United States agriculture; it has been implicated in the demasculinization of male frogs at concentrations as low as 1 ppb.

*Biomagnification* (also: *bioaccumulation*): Increasing potency and toxicity of POPs as they are consumed along a given food chain.

*Cancer Alley:* A nickname applied to a stretch of the Mississippi River between Baton Rouge and New Orleans, Louisiana, that includes several mostly low-income, mainly African-American communities in the shadows of petrochemical and plastics industries.

*Carcinogenic:* The propensity of a substance, such as an organochlorine chemical, to cause cancer in humans or animals.

*Chloracne:* Small pimples with dark pigmentation of the exposed area, followed by blackheads and pustules, induced by exposure to organochlorine compounds.

*Chlordane:* An organochlorine pesticide; one of the dirty dozen to be banned under the Stockholm Convention.

*Chlorine:* The basic chemical building block of the organochlorine chemical family. It is the seventeenth element on the Periodic Table. Chlorine's powerful disinfectant qualities stem from its ability to bond with and destroy the outer surfaces of bacteria and viruses.

*Chlorofluorocarbons (CFCs):* Organochlorine compounds widely used until the 1990s (under the trade name Freon) in air conditioning and other applications. During the 1980s, CFCs were found to be severely depleting stratospheric ozone levels. These compounds were banned by an international agreement (popularly known as the Montreal Protocol) negotiated during the late 1980s, but they still are manufactured clandestinely and smuggled by so-called Frio Banditos.

*DDT (Para-dichlorodiphenyltrichloroethane):* An organochlorine commonly used as a pesticide; one of the dirty dozen to be banned under the Stockholm Convention.

*Dibromochloropropane (DBCP):* An organochlorine soil fumigant that has caused infertility in men who have been exposed to the compound at their work sites.

*Dieldrin:* an organochlorine commonly used as a pesticide; one of the dirty dozen.

*Diethylstilbestrol (DES):* A form of synthetic estrogen that was given to at least five million women to prevent miscarriages and pregnancy complications. The chemical was banned in 1971 after it was linked to increases in a previously rare form of vaginal cancer.

*Dioxins:* A family of very toxic industrial chemicals to be banned under the Stockholm Protocol as part of the dirty dozen.

*Dirty dozen:* The twelve most commonly used toxic POPs slated for international ban under the Stockholm Convention. *See also* Aldrin, Chlordane, DDT, Dieldrin, Dioxins, Endrin, Furans, Heptachlor, Hexachlorobenzene, Mirex, PCBs, and Toxaphene.

*Dobson units:* A measurement of atmospheric ozone levels, most notably in the stratosphere.

*Endocrine disrupter:* A substance with the ability to interfere with hormone-related functions in animals associated, at different levels of contamination, with several of the organochlorines.

*Endrin:* An organochlorine pesticide; one of the dirty dozen to be banned under the Stockholm Convention.

*Estrogen:* A hormone that imparts feminine qualities.

*Estrogen mimicry:* The ability of organochlorine compounds to act with estrogenic properties in the bodies of humans and other animals and thereby interfere with reproduction. Such an effect by DDT was reported in chickens as early as 1950.

*Frio Banditos:* Smugglers of chlorofluorocarbons, notably Freon, usually from Mexico into the United States.

*Furans (a.k.a. polychlorinated dibenzofurans):* Industrial chemicals to be banned among the dirty dozen under the Stockholm Convention.

*Global warming:* The retention, near the Earth's surface, of heat provoked by rising levels, in the lower atmosphere, of heat-trapping trace gases, such as carbon dioxide, methane, and some of the organochlorines.

*Heptachlor:* One of the dirty dozen.

*Hexachlorobenzene:* An organochlorine pesticide; one of the dirty dozen to be banned under the Stockholm Convention.

*Intersex (also, imposex):* Having reproductive organs of both sexes (usually *intersex*), or secreting hormones of both sexes (usually *imposex*), a condition that may be induced in some animals (most notably some fish species and frogs) by exposure to organochlorine chemicals.

*Love Canal:* A neighborhood near Niagara Falls, New York, that was contaminated by a number of environmental toxic agents, including dioxins. In 1978 President Jimmy Carter declared Love Canal a federal disaster area, the first such declaration for a human-induced environmental disaster.

*Mercury Sunrise:* The first sunrise of Arctic spring that initiates a series of chemical reactions that dump mercury out of the atmosphere into the polar snow pack, from which it is released into the food web during the brief polar growing season. First reported in the Arctic during 1998.

*Methyl bromide:* A pesticide that is banned in several countries, including the Netherlands and Canada, but allowed in the United States as of this writing. This chemical may play a role in stratospheric ozone depletion.

*Mirex:* An organochlorine pesticide; one of the dirty dozen to be banned under the Stockholm Convention.

*Organochlorines:* Chlorine-based chemicals. *Organo,* in this case, refers to the presence of carbon in combination with chlorine.

*Ozone:* Three molecules of oxygen in combination that is depleted by chemical interaction in the stratosphere by some organochlorines. Ozone absorbs ultraviolet radiation that would otherwise contribute to skin cancers in human beings and other animals. *See also* Dobson units.

*Ozone hole:* Depletion of stratospheric ozone levels, notably over the Antarctic and Arctic, by chlorofluorocarbons and other chemical processes. The depletion does not form a hole in the ozone shield but depletes it to a point that allows dangerous levels of ultraviolet radiation to reach the surface.

*Persistent organic pollutants (POPs):* Carbon-based chemical compounds and mixtures that share four characteristics: high toxicity, persistence, a special affinity for fat, and a propensity to evaporate and travel long distances. These chemicals have been commonly used as herbicides and pesticides, or produced as by-products of several industrial processes, all of which release toxicity into the environment.

*Phytoremediation:* The use of plants to absorb heavy metals and organic chemicals.

*Polar stratospheric clouds (PSCs):* Earth's highest clouds, which form above the poles at altitudes of eighty-two to eighty-six kilometers. These clouds have been increasing in coverage and thickness during the last four decades, changes that may be linked to human emissions of carbon dioxide and methane, as well as to depletion of stratospheric ozone. Until recently, clouds at such heights were extremely rare.

*Polychlorinated biphenyls (PCBs):* A family of industrial chemicals; included among the dirty dozen to banned under the Stockholm Convention.

*Polyvinyl chloride (PVC):* A major constituent of many soft plastics, including toys (such as Barbie dolls); a major source, in manufacture, of several organochlorine pollutants, notably dioxins.

*Precautionary principle:* Developed at the Rio Earth Summit (1992), the precautionary principle holds that where there are threats of serious or irreversible damage, lack of full scientific certainty shall not be used as a reason for postponing cost-effective measures to prevent environmental degradation. The chemical industry has long argued that harm should be proven before products are removed, but environmentalists believe the search for a largely unattainable degree of scientific certainty is a corporate stalling tactic.

*Seveso, Italy:* Site of a massive dioxin spill during 1976.

*Silent Spring:* Title of a well-known book by Rachel Carson, published in 1962, which raised widespread popular concern about the effects of organochlorines (most notably DDT) on the environment.

*Sink:* An area, such as the Arctic or Antarctic, to which organochlorine compounds, most notably PCBs and dioxins, are carried by atmospheric and oceanic currents and then deposited. The word is used by climate scientists differently to indicate an area that absorbs carbon dioxide, as in a "carbon sink."

*Sperm count:* The number of motive (moving, or living) sperm in a given volume of a man's semen. Low sperm counts are said to be azoospermic (having no sperm to count) and oligospermic (having sperm counts of less than 20 million per milliliter).

*Stockholm Convention:* An international protocol that requires a ban of the twelve most commonly used persistent organic pollutants (the dirty dozen).

*Times Beach, Missouri:* A town that was contaminated by dioxins to a point where it became uninhabitable. The U.S. EPA bought the town, moved the residents out, and demolished its homes during 1983.

*Toxaphene:* One of the dirty dozen.

*Ultraviolet (UV) radiation:* Part of the sun's energy that is filtered out, for the most part, by stratospheric ozone. Depletion of ozone at this level allows UV radiation to reach the Earth, where it may cause skin cancer and other health problems. *See also* chlorofluorocarbons (CFCs).

# Selected Bibliography

"Activists Hail Recall of Toxic USA Mercury Shipment to India; Congress to Consider Storage and Sales/Export Ban." Press Release, Basil Action Network. January 29, 2001. http://www.ban.org.

"Additive Poses Hard Choice: Clean Air or Clean Water?" *Omaha World-Herald,* January 26, 2000.

"Africa: Fighting Back the Widening Deserts." BBC News, November 30, 1998. http://news.bbc.co.uk/hi/english/world/africa/newsid_224000/224597.stm.

Aiking H., M. B. van Acker, R. J. P. M. Scholten, J. F. Feenstra, and H. A. Valkenburg. "Swimming-Pool Chlorination: A Health Hazard?" *Toxicological Letters* 72 (1994): 375–80.

Alderdice, D. F., and M. E. Worthington. "Toxicity of a DDT Forest Spray to Young Salmon." *Canadian Fish Culturist* 24 (1959): 41–48.

Allen, Joe. "Malathion in Mission." *Circle* (Minneapolis), April 1995, 8–12.

Allen, W. "Dioxin Find Worries Residents: Many Have Questions about Chemical." *St. Louis Post-Dispatch,* August 10, 1997. http://lists.essential.org/1997/dioxin-l/msg00271.html.

Allen, W. "Dioxin Levels Found in Private Drive in Ellisville." *St. Louis Post-Dispatch,* August 6, 1997. http://lists.essential.org/1997/dioxin-l/msg00271.html.

Allsopp, Michelle, Pat Costner, and Paul Johnston. *Body of Evidence: The Effects of Chlorine on Human Health.* London: Greenpeace International, 1995. http://www.greenpeace.org/~toxics/reports/recipe.html.

Allsopp, Michelle, Ruth Stringer, and Paul Johnston. *Unseen Poisons: Levels of Organochlorine Chemicals in Human Tissue.* London: Greenpeace International, 1998.

Allsopp, Michelle, Ben Erry, Ruth Stringer, Paul Johnston, and David Santillo. "Recipe for Disaster: A Review of Persistent Organic Pollutants in Food." Research report. Exeter, U.K.: Greenpeace Research Laboratories and University of Exeter Department of Biology, 2000. http://www.greenpeace.org/~toxics/reports/recipe.html.

Ames, Bruce N., Margie Profet, and Lois Swirsky Gold. "Nature's Chemicals and Synthetic Chemicals: Comparative Toxicology." 1990. http://www.mapcruzin.com/environment21.

Anderson, J. G., W. H. Brune, and M. H. Proffitt. "Ozone Destruction by Chlorine Radicals within the Antarctic Vortex: The Spatial and Temporal Evolution of $ClO/O_3$, Anticorrelation Based on in Situ ER-2 Data." *Journal of Geophysical Research* 94 (1989): 11,465–79.

Anderson, Julie. "Diazinon Sales to Be Eased Out." *Omaha World-Herald,* December 6, 2000.

Ando, M., H. Saito, and I. Wakisaka. "Gas Chromatographic and Mass Spectrometric Analysis of Polychlorinated Biphenyls in Human Placenta and Cord Blood." *Environmental Research* 41 (1986): 14–22.

"Antarctic Ozone Hole Shrinks, Divides in Two." Environment News Service, September 30, 2002. http://ens-news.com/ens/sep2002/2002-09-30-03.asp.

Arnold, D. L., J. Mes, and F. Bryce. "A Pilot Study on the Effects of Aroclor 1254 Ingestion by Rhesus and Cynomolgus Monkeys As a Model for Human Ingestion of PCBs." *Food Chemistry and Toxicology* 28 (1990): 847–57.

Aronson, Richard B., William F. Precht, Ian G. MacIntyre, and Thaddeus J. T. Murdoch. "Coral Bleach-out in Belize." *Nature* 405 (May 4, 2000): 36.

"Atmospheric Science: Really High Clouds." *Science* 292 (April 13, 2001): 171.

Auger, J., J. M. Kuntsmann, F. Czyglik, and P. Jouannet. "Decline in Semen Quality among Fertile Men in Paris during the Past 20 Years." *New England Journal of Medicine* 332, no. 5 (1995): 281–85.

Auman, H. J., J. P. Ludwig, C. L. Summer, D. A. Verbrugge, K. L. Froese, T. Colborn, and J. P. Giesy. "PCBs, DDE, DDT, and TCDD-EQ in Two Species of Albatross on Sand Island, Midway Atoll, North Pacific." *Oceanography, Environment, Toxicological Chemistry* 16, no. 3 (1997): 498–504.

Austin, J., N. Butchart, and K. P. Shine. "Possibility of an Arctic Ozone Hole in a Doubled-$CO_2$ Climate." *Nature* 360 (November 19, 1992): 221–25.

Ayres, Ed. *God's Last Offer: Negotiating for a Sustainable Future.* New York: Four Walls Eight Windows, 1999.

"Bacteria Breaks Down PCBs in Baltimore Harbor." Environment News Service, January 8, 2002. http://ens-news.com/ens/jan2002/2002L-01-08-09.html.

Bae, J., E. L. Stuenkel, R. Loch-Caruso. "Stimulation of Oscillatory Uterine Contraction by the PCB Mixture Aroclor 1242 May Involve Increased

[Ca2 + ](i) through Voltage-Operated Calcium Channels." *Toxicology and Applied Pharmacology* 155, no. 3 (1999): 261–72.

Baker, Linda. "The Hole in the Sky: Think the Ozone Layer Is Yesterday's Issue? Think Again." *E: The Environmental Magazine,* November/December, 2000, 34–39. http://www.e-magazine.com/november-december 2000/1100feat2.html.

Barnett, J. B., T. Coldborn, and M. Fournier. "Consensus Statement from the Work Session on 'Chemically Induced Alterations in the Developing Immune System: The Human/Wildlife Connection:'" *Environmental Health Perspectives* 104, suppl. 4 (1996): 807–8.

Barsotti, D. A., R. J. Marlar, and J. R. Allen. "Reproductive Dysfunction in Rhesus Monkeys Exposed to Low Levels of Polychlorinated Biphenyls (Aroclor 1248)." *Food Chemistry and Toxicology* 14 (1976): 99–103.

Baughman, R., and M. Meselson. "An Analytical Method for Detecting TCDD (Dioxin): Levels of TCDD in Samples from Vietnam." *Environmental Health Perspectives* 5 (1973): 27–35.

Becher, H., K. Steindorf, and D. Flesch-Janys. "Quantitative Cancer Risk Assessment for Dioxin Using an Occupational Cohort." *Environmental Health Perspectives* 106 (1998): 663–70.

Beckmen, Kimberlee B., Margaret M. Krahn, and Jeffrey L. Stott. "Immunotoxicology of Organochlorine Contaminants in Northern Fur Seals, *Callorhinus Ursinus.*" Past Arctic Research Initiative Progress Reports: 1997. http://www.cifar.uaf.edu/proposal/award97/award97.html.

Beckmen, Kimberlee, Margaret M. Krahn, and Jeffrey L. Stott. "Toxicokinetics of Organochlorine Contaminants and Effects on Immune System Development in Free-ranging Northern Fur Seal Tissues." Past Arctic Initiative (ARI) Awards 1998–1999. http://www.cifar.uaf.edu/proposal/toxic.html.

Beek, Bernd. *The Handbook of Environmental Chemistry.* Berlin: Springer Verlag, 2001.

"Belgium Sees Dioxin Crisis Costing 60 Billion Belgian Francs," Reuters News Service, June 30, 1999.

Bernard, A., C. Hermans, F. Broeckaert, G. de Poorter, A. De Cock, and G. Houins. "Food Contamination by PCBs and Dioxins: An Isolated Episode in Belgium is Unlikely to Have Affected Public Health." *Nature* 401 (1999): 231–234.

Bernard, Harold W. Jr. *Global Warming: Signs to Watch For.* Bloomington: Indiana University Press, 1993.

Bertazzi, P. A., L. Riboldi, and A. Persatori. "Cancer Mortality of Capacitor Manufacturing Workers." *American Journal of Industrial Medicine* 11 (1987): 65–176.

Binder, Sarah. "United Nations Sets Out to Ban Chemicals like DDT, PCBs." *Ottawa Citizen,* June 29, 1998.

"Biomass Burning Boosts Stratospheric Moisture." Environment News Service, February 20, 2002. http://ens-news.com/ens/feb2002/2002L-02-20-09.html.

Blair, A., D. J. Grauman, J. H. Lubin, and J. F. Fraumeni. "Lung Cancer and Other Causes of Death among Licensed Pesticide Applicators." *Journal of the National Cancer Institute* 71 (1983): 31–37.

Blais, Jules M., David W. Schindler, Derek C. G. Muir, Lynda E. Kimpe, David B. Donald, and Bruno Rosenburg. "Accumulation of Persistent Organochlorine Compounds in Mountains of Western Canada." *Nature* 395 (October 8, 1998): 585–88.

Bloomfield, Janine. *Iowans Can Elect to Combat Climate Change Now.* Environmental Defense Fund. January 13, 2000. http://terra.whrc.org/links/links.htm.

Bodansky, Daniel. "The Precautionary Principle in U.S. Environmental Law." In *Interpreting the Precautionary Principle,* edited by Timothy O'Riordan and James Cameron. London: Earthscan Publications, 1994.

Bookchin, Murray [Lewis Herber]. *Our Synthetic Environment.* New York: Knopf, 1962.

Borenstein, Seth. "Arctic Lost 60 per cent of Ozone Layer; Global Warming Suspected." Knight-Ridder News Service, April 6, 2000.

Bowermaster, J. "A Town Called Morrisonville," *Audubon,* July–August 1993, 42–51.

Brasseur, Guy P., Anne K. Smith, Rashid Khosravi, Theresa Huang, Stacy Walters, Simon Chabrillat, and Gaston Kockarts. "Natural and Human-Induced Perturbations in the Middle Atmosphere: A Short Tutorial." In *Atmospheric Science across the Stratopause,* edited by David E. Siskind, Stephen D. Eckermann, and Michael E. Summers. Washington, D.C.: American Geophysical Union, 2000.

Briejer, C. J. "The Growing Resistance of Insects to Insecticides." *Atlantic Naturalist* 13, no. 3 (1958): 149–55.

British Broadcasting Corporation. "Severe Loss to Arctic Ozone." BBC News, April 5, 2000. http://news.bbc.co.uk/hi/english/sci/tech/news id_702000/702388.stm.

Brodkin, Marc, and Martin Simon, "The Effects of Aquatic Acidification on Rana Pipiens," *Froglog* 20 (January 1997): 3.

Brooks, Paul. *The House of Life: Rachel Carson at Work.* Boston: Houghton-Mifflin, 1972.

Brotons, J. A., M. F. Olea-Serrano, M. Villalobos, and N. Olea. "Xenoestrogens Released from Lacquer Coatings in Food Cans." *Environmental Health Perspectives* 102 (1995): 608–12.

Brown, David. "Defoliant Connected to Diabetes." March 29, 2000. http://irptc.unep.ch/pops/newlayout/press_items.htm.

Brown, DeNeen L. "Arctic Canada's Silent Invader: Contamination Threatens Native People's Way of Life in Fragile Region." *Washington Post,* May 17, 2001.

Brown, D. P. "Mortality of Workers Exposed to Polychlorinated Biphenyls: An Update." *Archives of Environmental Health* 42, no. 6 (1987): 333–39.

Bryant, D., L. Burke, J. McManus, and M. Spaulding. *Reefs at Risk: A Map-Based Indicator of Threats to the World's Coral Reefs.* Washington, D.C.: World Resources Institute, 1998.

"Bulletin Board: Vietnam Veterans Benefit from Agent Orange Rules." *Indian Country Today,* May 16, 2001.

Burlington, H., and V. F. Lindeman. "Effect of DDT on Testes and Secondary Sex Characteristics of White Leghorn Cockerels." *Proceedings of the Society for Experimental Biology and Medicine* 74 (1950): 48–51.

Cable News Network (CNN) Interactive. "Pesticides Suspected in Florida Gator Decline." March 15, 1998. http://www.cnn.com/EARTH/9803/15/gator.woes/index.html.

Cadbury, Deborah. *Altering Eden: The Feminization of Nature.* New York: St. Martin's Press, 1997.

Cadbury, Deborah. *The Estrogen Effect: How Chemical Pollution Is Threatening Our Survival.* New York: St. Martin's/Griffin, 2000.

Calamai, Peter. "Alert over Shrinking Ozone Layer." *Toronto Star,* March 18, 2002.

Calamai, Peter. "Chemical Fallout Hurts Inuit Babies." *Toronto Star,* March 22, 2000. http://irptc.unep.ch/pops/newlayout/press_items.htm.

Campagna, Darryl. "A Hudson Cleanup Faces Bitter Battle; General Electric Fights EPA Plan to Dredge River." *Boston Globe,* April 22, 2001.

Campbell, J. S., J. Wong, L. Wong, D. Tryphonas, D. Arnold, E. Nera, B. Cross, and B. LaBossiere (1985). "Is Simian Endometriosis an Effect of Immunotoxicity?" Paper presented at the Ontario Association of Pathologists 48th Annual Meeting, London, Ontario, 1985.

Canadian Polar Commission. *For Generations to Come: A Canadian Conference on Contaminants, the Environment, and Human Health in the Arctic, October 8–10, 1996.* Held at Iqaluit, Northwest Territories. http://www.polarcom.gc.ca/publict/reports/crjune97.html.

Cantor, D. S., G. Holder, W. Cantor, P. C. Kahn, G. C. Rodgers, G. H. Smoger, W. Swain, H. Berger, and S. Suffin. "*In-utero* and Postnatal Exposure to 2,3,7,8-TCDD in Times Beach, Missouri: Impact on Neurophysiological Functioning." Paper presented at Dioxin '93, 13th International Symposium on Chlorinated Dioxins and Related Compounds, Vienna, September 20–24, 1993.

Carlsen, E., A. Giwercman, N. Keiding, and N. E. Skakkebaek. "Evidence for Decreasing Quality of Semen during the Past 50 Years." *British Medical Journal* 305 (1992): 609–13.

Carson, Rachel. *Silent Spring*. Boston: Houghton-Mifflin, 1962.

Carson, Rachel. *Silent Spring*. Westport, Conn.: Fawcett Publications, 1962.

Case, R. A. M. "Toxic Effects of DDT in Man." *British Medical Journal* 1 (December 15, 1945): 842–45.

"Celebrities to Tour 'Cancer Alley,' Louisiana; Alice Walker, Alfred Woodard, and Mike Farrell among Speakers at National Town Meeting on Environmental Justice." N. d. http://www.greenpeaceusa.org/toxics/canceralleytour/celebritytour.htm.

"Chemical Exposure May Reduce Sperm Quality." Environment News Service, November 13, 2002. http://ens-news.com/ens/nov2002/2002-11-13-09.asp#anchor7.

Chen, P. H., K. T. Chang, and Y. D. Yu. "Polychlorinated Biphenyls and Polychlorinated Dibenzofurans in the Toxic Rice-bran Oil That Caused PCB Poisoning in Tai Chung." *Bulletin of Environmental Contamination and Toxicology* 26 (1981): 489–95.

Chen, Y.-C. J., Y.-L. Guo, and C.-C. Hsu. "Cognitive-Development of Yu-cheng (Oil Disease) Children Prenatally Exposed to Heat-Degraded PCBS." *Journal of the American Medical Association* 268 (1992): 3213–18.

Chilvers, C., M. C. Pike, and D. Foreman. "Apparent Doubling of Frequency of Undescended Testicles in England and Wales, 1962–1981." *Lancet* 2(8398) (1984): 330–32.

Chlorine Chemistry Council. Home Page. May 2001. http://www.c3.org/chlorine_what_is_it/chlorine_story2.html.

Chow, Gee. "Pesticides and the Mystery of Deformed Frogs," *Journal of Pesticide Reform* 17 (Fall 1997): 14.

Christie, Maureen. *The Ozone Layer: A Philosophy of Science Perspective.* Cambridge, U.K.: Cambridge University Press, 2001

Clausen, J., and O. Berg. "The Content of Polychlorinated Hydrocarbons in Arctic Ecosystems." *Pure and Applied Chemistry* 42 (1975): 223–26.

Cockell, Charles S., and Andrew R. Blaustein, eds. *Ecosystems, Evolution, and Ultraviolet Radiation.* New York: Springer, 2001.

Colborn, Theodora E., and Coralie Clement. "Chemically-Induced Alterations in Sexual and Functional Development: The Wildlife/Human Connection." In Theodora E. Colborn and Coralie Clement, eds. *Advances in Modern Environmental Toxicology*, vol. 21. Princeton, N.J.: Princeton Scientific Publishing, 1992.

Colborn, Theodora E., and Coralie Clement. "Statement from the Work Session on Chemically-Induced Alterations in Sexual and Functional Development: The Wildlife/Human Connection." In Theodora E. Colborn and Coralie Clement, eds. *Advances in Modern Environmental Toxicology*, vol. 21. Princeton, N.J.: Princeton Scientific Publishing, 1992.

Colborn, Theodora E., Alex Davidson, Sharon N. Green, R. A. (Tony) Hodge, C. Ian Jackson, and Richard A. Liroff. *Great Lakes/Great Legacy?* Washington, D.C.: Conservation Foundation, 1990.

Colborn, Theodora E., D. Dumanoski, and J. P. Myers. *Our Stolen Future: Are We Threatening Our Fertility, Intelligence, and Survival? A Scientific Detective Story*. New York: Penguin, 1996.

Colborn, Theodora E., Frederick S. vom Saal, and Ana M. Soto. "Developmental Effects of Endocrine-Disrupting Chemicals in Wildlife and Humans." *Environmental Health Perspectives* 101, no. 5 (1993): 378–84.

Colborn, Theodora E., and M. J. Smolen. "Epidemiological Analysis of Persistent Organochlorine Contaminants in Cetaceans." *Review of Environmental Contamination Toxicology* 146 (1996): 91–172.

Commission on the Assessment of Polychlorinated Biphenyls in the Environment. *Polychlorinated Biphenyls: A Report Prepared for the Environmental Studies Board, Commission on Natural Resources, National Research Council*. Washington, D.C.: National Research Council, 1979.

Committee on the Future Role of Pesticides in U.S. Agriculture, Board on Agriculture and Natural Resources and Board on Environmental Studies and Toxicology, Commission on Life Sciences, and National Research Council. *The Future Role of Pesticides in U.S. Agriculture*. Washington, D.C.: National Academies Press, 2000.

Committee on Toxicology. "Occupational Dieldrin Poisoning." *Journal of the American Medical Association* 172 (April 1960): 2077–80.

Commoner, Barry. *The Closing Circle: Nature, Man, and Technology*. New York: Knopf, 1971.

Commoner, Barry. *Making Peace with the Planet*. New York: Pantheon Books, 1990.

Commoner, Barry. *Science and Survival*. New York: Viking Press, 1966.

Cone, Marla. "Human Immune Systems May Be Pollution Victims." *Los Angeles Times*, May 13, 1996.

Connor, Steve. "Global Warming Is Blamed for First Collapse of a Caribbean Coral Reef." *London Independent*, May 4, 2000.

Connor, Steve. "Ozone Layer over Northern Hemisphere Is Being Destroyed at 'Unprecedented Rate.'" *London Independent*, March 5, 2000.

Cook, J. W., and E. C. Dodds. "Sex Hormones and Cancer-Producing Compounds." *Nature* (February 1933): 205–6.

Cook, Judith, and Chris Kaufman. *Portrait of a Poison: The 2,4,5-T Story*. London: Pluto Press, 1982.

Costner, Pat. *The Burning Question: Chlorine and Dioxin*. Washington, D.C.: Greenpeace USA, 1997.

Costner, Pat. "The Burning Question—Chlorine and Dioxin. Taking Back Our Stolen Future: Hormone Disruption and PVC Plastic." April 1997. http://www.greenpeace.org/~toxics/reports/tbosf/tbosf.html#Introduction.

Costner, Pat. *Dioxin Elimination: A Global Imperative.* Amsterdam, Neth.: Greenpeace International, 2000. www.greenpeace.org/~toxics under "reports." [Can also be found at http://www.chej.org.]

Costner, Pat. *PVC: A Primary Contributor to the U.S. Dioxin Burden; Comments Submitted to the U.S. EPA Dioxin Reassessment.* Washington, D.C. Greenpeace U.S.A., 1995.

Costner, Pat, C. Cray, G. Martin, B. Rice, D. Santillo, and R. Stringer. *PVC: A Principal Contributor to the U.S. Dioxin Burden.* Washington, D.C.: Greenpeace U.S.A., 1995.

Costner, Pat, D. Luscombe, and M. Simpson. *Technical Criteria for the Destruction of Stockpiled Persistent Organic Pollutants.* London: Greenpeace International, 1998.

Costner, Pat, and J. Thornton. *Playing with Fire: Hazardous Waste Incineration.* Washington, D.C.: Greenpeace U.S.A., 1991.

Coulston, F., and F. Pocchiari. *Accidental Exposure to Dioxins: Human Health Aspects.* New York: Academic Press, 1983.

Crain, D. A., L. J. Guillette, D. B. Pickford, H. F. Percival, and A. R. Woodward. "Sex-Steroid and Thyroid Hormone Concentrations in Juvenile Alligators (*Alligator Mississippiensis*) from Contaminated and Reference Lakes in Florida, USA." *Environmental Toxicology and Chemistry* 17, no. 3 (1998): 446–52.

Cray, Charlie. "Hundreds Oppose Shintech Proposal in Louisiana. Citizens and Other Interest Groups Cite Health Concerns." Greenpeace USA. December 9, 1996. http://lists.essential.org/1996/dioxin-l/msg 00752.html.

Cray, Charlie, and Monique Harden. "PVC and Dioxin: Enough Is Enough." *Rachel's Environment and Health Weekly* 616, September 18, 1998. http://csf.colorado.edu/envtecsoc/98/0285.html.

Crutzen, Paul J. "The Antarctic Ozone Hole, a Human-Caused Chemical Instability in the Stratosphere: What Should We Learn from It?" In *Geosphere–Biosphere Interactions and Climate,* edited by Lennart O. Bengtsson and Claus U. Hammer. Cambridge, U.K.: Cambridge University Press, 2001.

Dalton, Rex. "Frogs Put in the Gender Blender by America's Favourite Herbicide." *Nature* 416 (April 18, 2001): 665–66.

Danish Environmental Protection Agency. *Male Reproductive Health and Environmental Chemicals with Oestrogenic Effects.* Copenhagen: Danish Environmental Protection Agency, 1995.

Davis, D. L., H. L. Bradlow, M. Wolff, T. Woodruff, D. G. Hoel, and H. Anton-Culver. "Medical Hypothesis: Xenooestrogens As Preventable Causes of Breast Cancer." *Environmental Health Perspectives* 101, no. 5 (1993): 372–77.

Davis, Neil. *Permafrost: A Guide to Frozen Ground in Transition.* Fairbanks: University of Alaska Press, 2001.

DeFao, Janine. "Protesters Block Waste Operation; Three Arrested at Medical Incinerator." *San Francisco Chronicle,* August 8, 2001.

Dellinger, J. A., N. Kmiecek, and S. Gerstenberger. "Mercury Contamination of Fish in the Ojibwa Diet: 1. Walleye Fillets and Skin-on versus Skin-off Sampling." *Water, Air, and Soil Pollution* 80 (1995): 69–76.

Dennis, Guy, and Jonathan Leake. "Breast-feeding Mothers May Pass Toxins to Babies." *Times* (London), April 30, 2000.

DeVito, M. G., and L. S. Birnbaum. "Toxicology of Dioxin and Related Chemicals." In *Dioxins and Health,* edited by A. Schecter. New York: Plenum, 1994.

Dewailly, E., P. Ayotte, S. Bruneau, S. Gingras, M. Belles-Isles, and R. Roy. "Susceptibility to Infections and Immune Status in Inuit Infants Exposed to Organochlorines." *Environmental Health Perspectives* 108 (2000): 205–11.

Dewailly, E., P. Ayotte, S. Bruneau, C. Laliberte, D. C. G. Muir, and R. J. Nordstrom. "Human Exposure to Polychlorinated Biphenyls through the Aquatic Food Chain in the Arctic." *Organohalogen Compounds* 14 (1993): 173–76.

Dewailly, E., S. Bruneau, C. Laliberte, M. Belles-Illes, J. P. Weber, and R. Roy. "Breast Milk Contamination by PCB and PCDD/Fs in Arctic Quebec. Preliminary Results on the Immune Status of Inuit Infants." *Organohalogen Compounds* 13 (1993): 403–6.

Dewailly, E., S. Dodin, R. Verreault, P. Ayotte, L. Sauve, and J. Morin. "High Organochlorine Body Burdens in Women with Estrogen-Receptor Positive Breast Cancer." *Journal of the National Cancer Institute* 86 (1994): 232–34.

Dewailly, E., J. J. Ryan, C. Laliberte, S. Bruneau, J. P. Weber, S. Gingras, and G. Carrier. "Exposure of Remote Maritime Populations to Coplanar PCBs." *Environmental Health Perspectives* 102, suppl. 1 (1994): 205–9.

Diamond, E. "The Myth of the 'Pesticide Menace.'" *Saturday Evening Post,* September 21, 1963, 17–18.

Dicke, William. "Numerous U.S. Plant and Freshwater Species Found in Peril." *New York Times,* January 2, 1996.

Dieckmann, W. J., M. D. Davis, and R. E. Pottinger. "Does the Administration of DES during Pregnancy Have Any Therapeutic Value?" *American Journal of Obstetrics and Gynecology* 66 (1953): 1062–81.

"Dioxin Deception: How the Vinyl Industry Concealed Evidence of Its Dioxin Pollution." March 27, 2001. http://www.greenpeaceusa.org/toxics/dioxin_deceptiontext.htm.

"Dioxin Levels Still Up in Many Vietnamese." *Omaha World-Herald,* May 15, 2001.

Dobson, G. M. B., and D. N. Harrison. "Measurement of the Amount of Ozone in the Earth's Atmosphere, and Its Relation to Other Geophysical Conditions." *Proceedings of the Royal Society of London* A110 (1926): 660–93.

Dodds, E. C., L. Goldberg, and W. Lawson. "Oestrogenic Activity of Esters of Diethyl Stilboestrol." *Nature* (1938): 211–12.

Dodds, E. C., L. Goldberg, W. Lawson, and R. Robinson. "Oestrogenic Activity of Alkylated Stilboestrols." *Nature* (1938): 247–49.

Dodds, E. C., and W. Lawson. "Molecular Structure in Relation to Oestrogenic Activity: Compounds without a Phenanthrene Nucleus." *Proceedings of the Royal Society of London* 125, suppl. B (1938): 222–32.

Drinker, Cecil K., et al. "The Problem of Possible Systematic Effects from Certain Chlorinated Hydrocarbons." *Journal of Industrial Hygiene and Toxicology* 19 (September 1937): 283–311.

Duchin, Lelanie. "Greenpeace's Secret Sampling at U.S. Vinyl Plants: Dioxin Factories Exposed." April 1997. http://www.greenpeace.org/~toxics/reports/reports.html.

Dunlap, Thomas R. *DDT: Scientists, Citizens, and Public Policy.* Princeton, N.J.: Princeton University Press, 1981.

Dunnick, J. K., and R. L. Melnick. "Assessment of the Carcinogenic Potential of Chlorinated Water: Experimental Studies of Chlorine, Chloramine, and Trihalomethanes." *Journal of the National Cancer Institute* 85, no. 10 (1993): 817–23.

Durbin, Dick. "Graham Flunks the Cost-Benefit Test." *Washington Post,* July 16, 2001. http://www.washingtonpost.com/wp-dyn/articles/A1421-2001Jul15.html.

Durham, William, et al. "Insecticide Content of Diet and Body Fat of Alaskan Natives." *Science* 134, no. 3493 (1961): 1880–81.

Eaton, S. B., M. C. Pike, R. V. Short, N. C. Lee, J. Trussell, R. A. Hatcher, J. W. Wood, C. M. Worthman, N. G. Blurton Jones, M. J. Konner, K. R. Hill, R. Bailey, and A. M. Hurtado. "Women's Reproductive Cancers in Evolutionary Context." *Quarterly Review of Biology* 69, no. 3 (1994): 353–67.

Ebinghaus, Ralf, Hans H. Kock, Christian Temme, Jürgen W. Einax, Astrid G. Löwe, Andreas Richter, John P. Burrows, and William H. Schroeder. "Antarctic Springtime Depletion of Atmospheric Mercury." *Environmental Science and Technology* 36, no. 6 (March 15, 2002): 1238–44.

El-Bayoumy, K. "Environmental Carcinogens That May Be Involved in Human Breast Cancer Etiology." *Chemical Research in Toxicology* 5, 5 (1992): 585–90.

Elkington, J. *The Poisoned Womb: Human Reproduction in a Polluted World.* London: Harmondsworth, 1985.

Ell, Renate. "Bonn POPs Talks Fall Short of Expectations." *Nunatsiaq News,* March 31, 2000. http://www.nunatsiaq.com/archives/nunavut 000331/nvt20331_09.html.

"Environmental Estrogens Could Hamper Songbird Breeding." Environment News Service, May 29, 2002. http://ens-news.com/ens/may 2002/2002-05-29-09.asp.

Epstein, Samuel S. "Beware Carcinogens, Phthalates in Cosmetics." Environment News Service, July 15, 2002. http://ens-news.com/ens/jul2002/2002-07-15e.asp.

Erickson, Jim. "Boulder Team Sees Obstacle to Saving Ozone Layer; 'Rocks' in Arctic Clouds Hold Harmful Chemicals." *Rocky Mountain News* (Denver), February 9, 2001.

European Commission. 1997. *European Workshop on the Impact of Endocrine Disrupters on Human Health and Wildlife: Report of Proceedings, December 2–4, 1996, Weybridge, U.K.* Copenhagen, Denmark: European Commission. [Report EUR 17549.]

"European Parliament Votes for Substitution of PVC Plastic." Greenpeace International, April 3, 2001. www.greenpeace.org.

Evans, Marlene S. "Anthropogenic Activities and the Great Slave Lake Ecosystem." http://ecsask65.innovplace.saskatoon.sk.ca/pages/current/toxics/antact.htm.

Evers, E. *De Vorming van PCDFs, PCDDs en Gerelateerde Verbindingen bij de Oxychlorering van Etheen. Vakgroep Milieu en Toxicologische Chemie.* Report MTC89EE, University of Amsterdam, Amsterdam, Neth., 1989.

Evers, E. H. G., R. W. P. M. Laane, G. J. J. Groeneveld, and Olie K. Levels. "Temporary Trends and Risk of Dioxin and Related Compounds in the Dutch Aquatic Environment." *Organohalogen Compounds* 28 (1996): 117–22.

Fahey, D. W., R. S. Gao, K. S. Carslaw, J. Kettleborough, P. J. Popp, M. J. Northway, J. C. Holecek, S. C. Ciciora, R. J. McLaughlin, T. L. Thompson, R. H. Winkler, D. G. Baumgardner, B. Gandrud, P. O. Wennberg, S. Dhaniyala, K. McKinney, T. Peter, R. J. Salawitch, T. P. Bui, J. W. Elkins, C. R. Webster, E. L. Atlas, H. Jost, J. C. Wilson, R. L. Herman, A. Kleinböhl, and M. von König. "The Detection of Large $HNO_3$-Containing Particles in the Winter Arctic Stratosphere." *Science* 291 (February 9, 2001): 1026–31.

Farman, J. C., B. G. Gardiner, and J. D. Shanklin. "Large Losses of Total Ozone Reveal Seasonal ClOx/NOx Interaction." *Nature* 315 (1985): 207–10.

Fein, G. G., Joseph L. Jacobson, and Sandra W. Jacobson. "Prenatal Exposure to Polychlorinated Biphenyls: Effects on Birth Size and Gestational Age." *Journal of Pediatrics* 105 (1984): 315–20.

Feldman, D., L. G. Tokes, P. A. Stathis, and D. Harvey. "Identification of 17B-oestradiol As the Estrogenic Substance in Saccharmyces Cerevisae." *Proceedings of the National Academy of Sciences* 81 (1984): 4722–28.

Fingerhut, M. A., W. E. Halperin, D. A. Marlow, L. A. Piacitelli, P. A. Honchar, M. H. Sweeney, A. L. Greife, P. A. Dill, K. Steenland, and A. J. Suruda. "Cancer Mortality in Workers Exposed to 2,3,7,8-tetrachlorodibenzo-p-dioxin." *New England Journal of Medicine* 324 (1991): 212–18.

Fitzgerald, E. F., K. A. Brix, D. A. Deres, S. A. Hwang, B. Bush, G. L. Lambert, and A. Tarbell. "Polychlorinated Biphenyl (PCB) and Dichlorodiphenyl Dichloroethylene (DDE) Exposures among Native American Men from Contaminated Great Lakes Fish and Wildlife." *Toxicology and Industrial Health* 12 (1996): 361–68.

Fitzgerald, E. F., S. Hwang, K. A. Brix, B. Bush, J. Quinn, and K. Cook. "Exposure to PCBs from Hazardous Waste among Mohawk Women and Infants at Akwesasne." Report for the Agency for Toxic Substances and Disease Registry, Atlanta, 1995.

Flanery, James Allen. "Debate on Water Re-ignites: Herbicides and Risk of Cancer Reported." *Omaha World-Herald*, October 19, 1994.

Flesch-Janys, Dieter, Jurgen Berger, Petra Gurn, Alfred Manz, Sibylle Nagel, Hiltraud Waltsgott, and James H. Dwyer. "Exposure to Polychlorinated Dioxins and Furans (PCDD/F) and Mortality in a Cohort of Workers from a Herbicide-Producing Plant in Hamburg, Federal Republic of Germany." *American Journal of Epidemiology* 142 (1995): 1165–75.

Flesch-Janys, D., J. Steindorf, P. Gurn, and H. Becher. "Estimation of the Cumulated Exposure to Polychlorinated Dibenzo-p-dioxins/furans and Standardized Mortality Ratio Analysis of Cancer Mortality by Dose in an Occupationally Exposed Cohort." *Environmental Health Perspectives* 106 (1998): 655–62.

Folmar, Leroy C., et al. "Vitellogenin Induction and Reduced Serum Testosterone Concentrations in Feral Male Carp (*Cyprinus Carpio*) Captured near a Major Metropolitan Sewage Treatment Plant." *Environmental Health Perspectives* 104 (1996): 1096–101.

Forberg, S., O. Tjelvar, and M. Olsson "Radiocesium in Muscle Tissue of Reindeer and Pike from Northern Sweden before and after the Chernobyl Accident: A Retrospective Study on Tissue Samples from the Swedish Environmental Specimen Bank." *Science of the Total Environment* 115 (1991): 179–89.

Foster, Krishna L., Robert A. Plastridge, Jan W. Bottenheim, Paul B. Shepson, Barbara J. Finlayson-Pitts, and Chester W. Spicer. "The Role of $Br_2$ and BrCl in Surface Ozone Destruction at Polar Sunrise." *Science* 291 (January 19, 2001): 471–74.

Franz, Neil. "EPA Sets Course to Complete Dioxin Reassessment." *Chemical Week*, June 21, 2000, 18.

Freeman, James. "Ozone Repair Could Bring New Problem." *Glasgow* (Scotland) *Herald*, April 25, 2001.

Freeman, Milton M. R. *Endangered Peoples of the Arctic: Struggles to Survive and Thrive.* Westport, Conn.: Greenwood Press, 2000.

Fry, D., C. Roone, S. Speich, and R. Peard. "Sex-Ratio Skew and Breeding Patterns of Gulls: Demographic and Toxicological Considerations." *Studies in Avian Biology* 10 (1987): 26–43.

Fry, D. M. "Reproductive Effects in Birds Exposed to Pesticides and Industrial Chemicals." *Environmental Health Perspectives* 103, suppl. 7 (1995): 1165–72.

Fry, D. M., and C. K. Toone. "DDT-Induced Feminization of Gull Embryos." *Science* 213 (1981): 922–24.

Furst, P., C. Furst, and K. Wilmers. "Human Milk As a Bio-indicator for Body Burden of PCDDs, PCDFs, Organochlorine Pesticides, and PCBs." *Environmental Health Perspectives: Supplement* 102, suppl. 1 (1994): 187–93.

Gannon, N., et al. "Storage of Dieldrin in Tissues and Its Excretion in Milk of Diary Cows Fed Dieldrin in Their Diets." *Journal of Agriculture and Food Chemistry* 7, no. 12 (1959): 824–32.

Gardner, Michael. "Gas Refiner Replaces Disputed Additive; MTBE Is Out, Ethanol Soon to Be In for Tosco." *San Diego Union,* December 22, 2000.

Geisy, J. P., J. Ludwig, and D. E. Tillitt. "Deformities in Birds of the Great Lakes Region." *Environmental Science and Technology* 28, no. 3 (1994): 128–35.

"Giant Canadian Food Chain Rejects Chemical Pesticides." Environment News Service, March 12, 2002. http://ens-news.com/ens/mar2002/2002L-03-12-01.html.

Gibbs, Lois Marie. *Dying from Dioxin: A Citizen's Guide to Reclaiming Our Health and Rebuilding Democracy.* Boston: South End Press, 1995.

Giwercman, A., and N. E. Skakkebaek. "The Human Testis: An Organ at Risk?" *International Journal of Andrology* 15 (1992): 373–75.

Goldey, E. S., L. S. Kehn, C. Lau, G. L. Rehnberg, and K. M. Crofton. "Developmental Exposure to Polychlorinated Biphenyls (Aroclor 1254) Reduces Circulating Thyroid Hormone Concentrations and Causes Hearing Deficits in Rats." *Toxicology and Applied Pharmacology* 135 (1998): 77–88.

Gordon, Anita. "New Report Concludes Nation Is Awash in Chemicals That Can Affect Child Development and Learning: Louisiana, Texas Emissions Lead the Country in First Effort Ever to Assess Scope and Sources of Developmental and Neurological Toxin Pollution; Report Documents Disturbing Trends in Developmental and Learning Deficits." Press release, September 7, 2000. Washington, D.C.: Physicians for Social Responsibility. http://www.psr.org/trireport.html.

Goreau, Thomas J., Raymond L. Hayes, Jenifer W. Clark, Daniel J. Basta, and Craig N. Robertson. "Elevated Sea-Surface Temperatures Cor-

266 Selected Bibliography

relate with Caribbean Coral Reef Beaching." In *A Global Warming Forum: Scientific, Economic, and Legal Overview,* edited by Richard A. Geyer. Boca Raton, Fla.: CRC Press, 1993, 225–262.

Goulden, M. L., S. C. Wofsy, J. W. Harden, S. E. Trumbone, P. M. Crill, S. T. Gower, T. Fries, B. C. Daube, S.-M. Fan, D. J. Sutton, A. Bazzaz, and J. W. Munger. "Sensitivity of Boreal Forest Carbon Dioxide to Soil Thaw." *Science* 279 (January 9, 1998): 214–17.

Graham, Frank Jr. *Since Silent Spring.* Boston: Houghton-Mifflin, 1970.

Gray, L. E. Jr., W. R. Kelce, E. Monosson, J. S. Ostby, and L. S. Birnbaum. "Exposure to TCDD during Development Permanently Alters Reproductive Function in Male Long Evans Rats and Hamsters: Reduced Ejaculated and Epididymal Sperm Numbers and Sex Accessory Gland Weights in Offspring with Normal Androgenic Status." *Toxicology and Applied Pharmacology* 131 (1995): 108–18.

Green, Emily. "Common Weed Killer Causes Sexual Abnormalities in Frogs, Study Claims." *Los Angeles Times,* April 16, 2002.

"Greens Oppose US Scheme to Dump Toxic Used Mercury in India." Greenpeace press release, December 2000.

"Group Attacks Water Quality in the Midwest." *Omaha World-Herald,* August 12, 1997.

Guillette, L. J., T. S. Gross, G. R. Masson, J. M. Matter, H. F. Percival, and A. R. Woodward. "Developmental Abnormalities of the Gonad and Abnormal Sex Hormone Concentrations in Juvenile Alligators from Contaminated and Control Lakes in Florida." *Environmental Health Perspectives* 102, no. 9 (1994): 680–88.

Guillette, L. J., T. S. Gross, D. A. Gross, A. A. Rooey, and H. F. Percival. "Gonad Steroidogenesis in Vitro from Juvenile Alligators Obtained from Contaminated Control Lakes." *Environmental Health Perspectives* 103, suppl. 4 (1995): 31–36.

Guillette, L. J., M. M. Meza, M. G. Aquilar, A. D. Soto, and I. E. Garcia. "An Anthropological Approach to the Evaluation of Preschool Children Exposed to Pesticides in Mexico." *Environmental Health Perspectives* 106, no. 6 (1998): 347–53.

Guo Y. L., Y. C. Chen, M. L. Yu, and C. H. Chen. "Early Development of Yu-Cheng Children Born Seven to Twelve Years after the Taiwan PCB Outbreak." *Chemosphere* 29, nos. 9–11 (1994): 2395–404.

Guo, Y. L., T. J. Lai, S. H. Ju, Y. C. Chen, and C. C. Hsu. "Sexual Development and Biological Findings in Yucheng Children." *Organohalogen Compounds* 14 (1993): 235–38.

Halliday, Tim. "1996 International Union for Conservation of Nature Red List." *Froglog* 21 (March 1997): 2. http://acs-info.open.ac.uk/info/newsletters/FROGLOG.html.

Handyside, Gillian. "New Dioxin Food Scare Strikes Belgium." Reuters News Service, August 4, 1999. http://platon.ee.duth.gr/data/maillist-archives/oikologia/1998-9/msg00337.html.

Harada, M. 1976. "Intrauterine Poisoning: Clinical and Epidemiological Studies and Significance of the Problem." *Bulletin of the Institute of Constitutional Medicine,* Supplement to vol. 25. (Kumamato University, Japan): 169–84.

Hardell, L., M. Ericksson, O. Axelson, and S. H. Zahm. "Cancer Epidemiology." In *Dioxins and Health,* edited by A. J. Schecter. New York: Plenum, 1994.

Harrington, R. W., and W. L. Bidlingmayer. "Effects of Dieldrin on Fishes and Invertebrates of a Salt Marsh." *Journal of Wildlife Management* 22, no. 1 (1958): 76–82.

Harris, J. R., M. E. Lippman, U. Veronesi, and W. Willett. "Breast Cancer (First of Three Parts)." *New England Journal of Medicine* 327, no. 5 (1992): 319–28.

Hartmann, Dennis L., John M. Wallace, Varavut Limpasuvan, David W. J. Thompson, and James R. Holton. "Can Ozone Depletion and Global Warming Interact to Produce Rapid Climate Change?" *Proceedings of the National Academy of Sciences* 97, no. 4 (February 15, 2000): 1412–17.

Hayes, Tyrone B., Atif Collins, Melissa Lee, Magdelena Mendoza, Nigel Noriega, A. Ali Stuart, and Aaron Vonk. "Hermaphroditic, Demasculinized Frogs after Exposure to the Herbicide Atrazine at Low Ecologically Relevant Doses." *Proceedings of the National Academy of Sciences* 99, no. 8 (April 16, 2002): 5476–80.

Hayes, Wayland J. Jr. "The Toxicity of Dieldrin to Man." *Bulletin of the World Heath Organization* 20 (1959): 891–912.

Hayland, Wayland J. Jr. et al. "Storage of DDT and DDE in People with Different Degrees of Exposure to DDT." *American Medical Association Archives of Industrial Health* 18 (November 1958): 398–406.

Hayona, Singy. "Zambia Struggles to Control Toxic PCBs." Environment News Service, September 1, 2000. http://www.repp.org/discussion/stoves/20009/ms0001.html.

"Heavy Metal Levels in Reindeer, Caribou, and Plants of the Seward Peninsula." Report, Reindeer Research Program, University of Alaska at Fairbanks. April 2000. http://reindeer.salrm.alaska.edu/research.htm.

Henderson, Mark. "Ozone Hole Will Heal in 50 Years, Say Scientists." *Times* (London), December 4, 2000.

Henry S., G. Cramer, M. Bolger, J. Springer, and R. Scheuplein. "Exposures and Risks of Dioxin in the U.S. Food Supply." *Chemosphere* 25, nos. 1–2 (1992): 235–38.

Herbst, A. L., H. Ulfelder, and D. C. Peskanzer. "Adenocarcinoma of the Vagina: Association of Maternal Stilboestrol Therapy with Tumor Appearances in Young Women." *New England Journal of Medicine* 284 (1971): 878–81.

Hill, Miriam. "Iqaluit's Waste Woes Won't Go Away; City Sets Up Bins Where Residents Can Dump Plastics, Metal." *Nunatsiaq News*, July 27, 2001.

Hines, M. "Surrounded by Oestrogens? Considerations for Neurobehavioural Development in Human Beings." In *Chemically-Induced Alterations in Sexual and Functional Development: The Wildlife/Human Connection*, edited by T. Colborn and C. Clement. Princeton, N.J.: Princeton Scientific Publishing, 1992.

Hogarth, Murray. "Sea-Warming Threatens Coral Reefs." *Sydney* (Australia) *Morning Herald*, November 26, 1998. http://www.smh.com.au/news/9811/26/text/national13.html.

Holden, A. V., and K. Marsden. "Organochlorine Pesticides in Seals and Porpoises." *Nature* 216 (1967): 1275–76.

Holloway, Marguerite. "Dioxin Indictment." *Scientific American* 270 (January 1994): 25.

Hong, R., K. Taylor, R. Abanour. "Immune Abnormalities Associated with Chronic TCDD Exposure in Rhesus." *Chemosphere* 18 (1989): 313–20.

Howdeshell, K. L., A. K. Hotchkiss, and Frederick vom Saal. "Exposure to Bisphenol A Advances Puberty." *Nature* 401 (1999): 763–64.

Høyer, A. P., P. Grandjean, T. Jørgensen, J. W. Brock, and H. B. Hartvig. "Organochlorine Exposure and Risk of Breast Cancer." *Lancet* 352 (1998): 1816–20.

Høyer, A. P., T. Jørgensen, J. W. Brock, and P. Grandjean. "Organochlorine Exposure and Breast Cancer Survival." *Journal of Clinical Epidemiology* 53 (2000): 323–30.

Hsu, S. T., C. I. Ma, S. K. Hsu, S. S. Wu, N. H. M. Hsu, C. C. Yeh, and S. B. Wu. "Discovery and Epidemiology of PCB Poisoning in Taiwan: A Four-Year Follow-up." *Environmental Health Perspectives* 59 (1985): 5–10.

Huff, J. "Dioxins and Mammalian Carcogenesis." In *Dioxins and Health*, edited by A. J. Schecter. New York: Plenum, 1994.

Huhndorf, Shari M. *Going Native: Indians in the American Cultural Imagination*. Ithaca, N.Y.: Cornell University Press, 2001.

Humphrey, H. E. B. "Population Studies of PCBs in Michigan Residents." In *PCBs: Human and Environmental Hazards*, edited by F. M. D'Itri and M. Kamrin. Boston, Mass.: Butterworth, 1983.

Hynes, H. Patricia. *The Recurring Silent Spring*. New York: Pergamon Press, 1989.

"Indian Enviros Urge Ban on Pesticide Endosulfan." Environment News Service, July 3, 2002. http://ens-news.com/ens/jul2002/2002-07-03-02.asp.

Indigenous Environmental Network, "Indigenous Peoples and POPs." Briefing paper for INC-4, February 2000. http://www.alphacdc.com/ien/pops_bonn_ien11.html.

Institute of Medicine, Committee to Review Health Effects in Vietnam Veterans of Exposure to Herbicides. *Veterans and Agent Orange.* Washington, D.C.: National Academies Press, 1994.

International POPs Elimination Network (IPEN). *Background Statement/ Platform.* http://www.psr.org/ipen/platform.htm.

Irvine, D. S. "Falling Sperm Quality." *British Medical Journal* 309 (1994): 476–79.

Irvine, S., E. Cawood, D. Richardson, E. MacDonald, and J. Aitken. "Evidence for Deteriorating Semen Quality in the United Kingdom: Birth Cohort Study in 577 Men in Scotland over 11 Years." *British Medical Journal* 312 (1996): 467–71.

Jackson, M. B., C. Chilvers, and M. C. Pike. "Cryptorchidism: An Apparent Substantial Increase since 1960." *British Medical Journal* 293 (1986): 1401–4.

Jacobson, Joseph L., G. G. Fein, Sandra W. Jacobson, P. M. Schwartz, and J. K. Dowler. "The Effect of Interuterine PCB Exposure on Visual Recognition Memory." *Child Development* 56 (1985): 853–60.

Jacobson, Joseph L., and Sandra W. Jacobson. "Effects of Exposure to PCBs and Related Compounds on Growth and Activity in Children." *Neurotoxicology and Teratology* 12 (1990): 319–26.

Jacobson, Joseph L., and Sandra W. Jacobson. "A Four-Year Follow-up Study of Children Born to Consumers of Lake Michigan Fish." *Journal of Great Lakes Research* 19, no. 4 (1993): 776–83.

Jacobson, Joseph L., and Sandra W. Jacobson. "Intellectual Impairment in Children Exposed to Polychlorinated Biphenyls *in Utero.*" *New England Journal of Medicine* 335, no. 11 (1996): 783–89.

Jacobson, Joseph L., Sandra W. Jacobson, and H. E. B. Humphrey. "Effects of *in Utero* Exposure to Polychlorinated Biphenyls and Related Contaminants on Cognitive Functioning in Young Children." *Journal of Pediatrics* 116 (1990): 38–45.

Jacobson, Joseph W., Sandra W. Jacobson, R. J. Padgett, G. A. Brumitt, and R. L. Billings. "Effects of Prenatal PCB Exposure on Cognitive Processing Efficiency and Sustained Attention." *Developmental Psychology* 28 (1992): 297–306.

Jacobson, Sandra W., G. G. Fein, and Joseph L. Jacobson. "The Effect of Intrauterine PCB Exposure on Visual Recognition Memory." *Child Development* 56 (1985): 856–60.

Jefferies, D. J. "The Role of the Thyroid in the Production of Sublethal Effects by Organochlorine Insecticides and Polychlorinated Biphenyls." In F. Moriarty, ed. *Organochlorine Insecticides: Persistent Organic Pollutants.* London: Academic Press.

Johansen, Bruce E. "Arctic Heat Wave." *Progressive,* October 2001, 18–20.

———. "Ecomania at Home; Ecocide Abroad." *University of Washington Daily,* May 24, 1972.

————. *The Global Warming Desk Reference.* Westport, CT: Greenwood Press, 2001.

————. *Life and Death on Mohawk Country.* Golden, Colo.: North American Press/Fulcrum, 1993.

————. "Pristine No More: The Arctic, Where Mother's Milk Is Toxic." *Progressive,* December 2000, 27–29.

Johnson, B. L., H. E. Hicks, D. E. Jones, W. Cibulas, A. Wargo, and C. T. De Rosa. "Public Health Implications of Persistent Toxic Substances in the Great Lakes and St. Lawrence Basins." *Journal of Great Lakes Research* 24, no. 2 (1998): 698–722.

Johnston, P., and I. McCrea, eds. *Death in Small Doses.* London: Greenpeace International, 1992.

Jones, K. C., and Y. Samiullah. "Deer Antlers As Pollution Monitors in the United Kingdom." *Deer* 6 (1985): 253–55.

Jucks, K. W., and R. J. Salawitch. "Future Changes in Atmospheric Ozone." In *Atmospheric Science across the Stratosphere,* edited by David E. Siskind, Stephen D. Eckermann, and Michael E. Summers. Washington, D.C.: American Geophysical Union, 2000.

"Jury Labels MTBE Gasoline As Defective Product." Environment News Service, April 18, 2002. http://ens-news.com/ens/apr2002/2002L-04-18-09.html#anchor3.

Kahn, P. C., M. Gochfeld, M. Nygren, M. Hansson, C. Rappe, H. Velez, T. Ghent-Guenther, and W. P. Wilson. "Dioxins and Dibenzofurans in Blood and Adipose Tissue of Agent Orange-Exposed Vietnam Veterans and Matched Controls." *Journal of the American Medical Association* 259 (1988): 1661–67.

Kaiser, Jocelyn, and Martin Enserink. "Treaty Takes a POP at the Dirty Dozen." *Science* 290 (December 15, 2000): 2053.

Kamrin, Michael A., and Paul W. Ridgers. *Dioxins in the Environment.* Washington, D.C.: Hemisphere Publishing, 1985.

Kazman, Sam, Eric Askanase, and Julie DeFalco. "A Petition to Declare Times Beach, Missouri, a National Historic Landmark." 1996. http://www.cei.org/MonoReader.asp?ID=518.

Keenleyside, M. H. A. "Insecticides and Wildlife." *Canadian Audubon* 21, no. 1 (1950): 1–7.

Kerr, Richard A. "Deep Chill Triggers Record Ozone Hole." *Science* 282 (October 16, 1998): 391.

————. "Stratospheric 'Rocks' May Bode Ill for Ozone." *Science* 291 (February 9, 2001): 962–63.

Kidd, Karen A., David W. Schindler, Raymond H. Hesslein, and Derek C. G. Muir. "Effects of Trophic Position and Lipid on Organochlorine Concentrations in Fishes from Subarctic Lakes in Yukon Territory." *Canadian Journal of Fisheries and Aquatic Sciences* 55, no. 4 (April 1998): 869–81.

Kirby, Alex. "Costing the Earth." British Broadcasting Corporation, Radio Four, October 26, 2000. http://news.bbc.co.uk/hi/english/sci/tech/ newsid_990000/990391.stm.

———. "Scientists Test Sex-Change Bears." British Broadcasting Company News, September 1, 2000. http://irptc.unep.ch/pops/newlayout/ press_items.htm.

Kolata, Gina. "PCB Exposure Linked to Birth Defects in Taiwan." *New York Times,* August 2, 1988.

Koopman-Esseboom, C., D. C. Morse, N. Weisglas-Kuperus, I. J. Lutkes-chipholt, C. G. Van der Paauw, L. G. M. T. Tuinstra, A. Brouwer, and P. J. J. Sauer. "Effects of Dioxins and Polychlorinated Biphenyls on Thyroid Hormone Status of Pregnant Women and Their Infants." *Pediatric Research* 36, no. 4 (1994): 468–73.

Koopman-Esseboom C., M. Huisman, N. Weisglas-Kuperus, C. G. van der Paauw, L. G. M. T. Tunistra, E. R. Boersma, and P. J. J. Sauer. "PCB and Dioxin Levels in Plasma and Human Breast Milk of 418 Dutch Women and Their Infants: Predictive Value of PCB Congener Levels in Maternal Plasma for Fetal and Infant's Exposure to PCBs and Dioxins." *Chemosphere* 28, no. 9 (1994): 1721–32.

Koopman-Esseboom, C., D. C. Morse, N. Weisglas-Kuperus, I. J. Lutkes-chipholt, C. G. Van der Paauw, L. G. M. T. Tuinstra, A. Brouwer, and P. J. J. Sauer. "Effects of Dioxins and Polychlorinated Biphenyls on Thyroid Hormone Status of Pregnant Women and Their Infants." *Pediatric Research* 36, no. 4 (1994): 468–73.

Krajick, Kevin. "Arctic Life, on Thin Ice." *Science* 291 (January 19, 2001): 424–25.

Kreiss, K., M. M. Zack, and R. D. Kimbrough. "Association of Blood Pressure and Polychlorinated Biphenyl Levels." *Journal of the American Medical Association* 245 (1981): 2505–9.

Krimsky, Sheldon. *Hormonal Chaos: The Scientific and Social Origins of the Environmental Endocrine Hypothesis.* Baltimore: Johns Hopkins University Press, 1999.

Krook, L., and G. A. Maylin. "Industrial Fluoride Pollution: Chronic Fluoride Poisoning in Cornwall Island Cattle." *Cornell Veterinarian* 69, suppl. 8 (1979): 1–70. http://www.ncbi.nlm.nih.gov/htbin-post/Entrez/ query?uid = 467082&form = 6&db = m&Dopt = r.

Kuratsune, M., M. Ikeda, Y. Nakamura, and T. Hirohata. "A Cohort Study on Mortality of Yusho Patients: A Preliminary Report." In *Unusual Occurrences As Clues to Cancer Etiology,* edited by R. W. Miller. Tokyo: Japan Scientific Society Press/Taylor and Francis, 1988.

Kurtz, Howard. "Moyers's Exclusive Report: Chemical Industry Left Out." *Washington Post,* March 22, 2001.

"Labeling Cosmetics May Help Prevent Cancers." Environment News Service, August 15, 2002. http://ens-news.com/ens/aug2002/2002-08-15-01.asp.

"*The Lancet* Press Release: Dioxin Exposure Linked to Long-term Decrease in Male Births." *Lancet,* May 27, 2000. http://irptc.unep.ch/pops/new layout/press_items.htm.

Lassek, E., D. Jahr, and R. Mayer. "Polychlorinated Dibenzo-p-dioxins and Dibenzofurans in Cow's Milk from Bavaria." *Chemosphere* 27, no. 4 (1993): 519–34.

Laurance, Jeremy. "Incinerator Pollution Can Have Devastating Effect on Birth Rate." *London Independent,* May 26, 2000. http://irptc.unep.ch/pops/newlayout/press_items.htm.

LaVecchio, F. A., H. M. Pashayan, and W. Singer. "Agent Orange and Birth Defects." *New England Journal of Medicine* 308 (1983): 719–20.

Lazaroff, Cat. "Beauty Products May Contain Controversial Chemicals." Environment News Service, July 10, 2002. http://ens-news.com/ens/jul2002/2002-07-10-07.asp.

———. "EPA Authorizes Hudson River Cleanup." Environment News Service. December 5, 2001. http://ens-news.com/ens/dec2001/2001L-12-05-06.html.

Lean, Geoffrey. "Poison Saves Hunted Whales." *London Independent,* January 9, 2000. http://irptc.unep.ch/pops/newlayout/press_items.htm.

———. "Quarter of World's Corals Destroyed." *London Independent,* January 7, 2001.

———. "World Industry Poisons Arctic Purity; A Climatic Trick Dumps Chemicals from Afar on People and Animals in the Far North." *London Independent,* December 15, 1996, 15.

Lear, Linda. *Rachel Carson: Witness for Nature.* New York: Henry Holt, 1997.

Liem, A. K. D., R. Hoogerbrugge, P. R. Koostra, E. G. Van der Velde, and A. P. J. M. De Jong. "Occurrence of Dioxins in Cow's Milk in the Vicinity of Municipal Waste Incinerators and a Metal Reclamation Plant in the Netherlands." Chemosphere 23 (1991): 1675–84.

"Links Found to Frog Deformities." Associated Press, July 8, 2002. http://www.cnn.com/2002/US/07/08/deformed.frogs.ap/index.html.

Lindberg, Steve E., Steve Brooks, C.-J. Lin, Karen J. Scott, Matthew S. Landis, Robert K. Stevens, Mike Goodsite, and Andreas Richter. "Dynamic Oxidation of Gaseous Mercury in the Arctic Troposphere at Polar Sunrise." *Environmental Science and Technology* 36, no. 6 (March 15, 2002): 1245–56.

Linden, Eugene. "The Big Meltdown: As the Temperature Rises in the Arctic, It Sends a Chill around the Planet." *Time,* September 4, 2000, 52.

Lipnick, Robert L., Joop L. M. Hermens, Kevin Jones, and Derek C. G. Muir. *Persistent, Bioaccumulative, and Toxic Chemicals.* Vol. 1, *Fate and Exposure.* New York: Oxford University Press, 2001.

Lipson, A. "Agent Orange and Birth Defects" *New England Journal of Medicine* 309 (1983): 491–95.

Lok, Corie, and Douglas Powell. "The Belgian Dioxin Crisis of the Summer of 1999: A Case Study in Crisis Communications and Management." February 1, 2000. http://www.plant.uoguelph.ca/riskcomm/crisis/belgian-dioxin-crisis-feb01-00.

Longgood, William F. *The Poisons in Your Food*. New York: Simon and Schuster, 1960.

Lonky, E., J. Reihman, T. Darvill, J. Mather, and H. Daly. "Neonatal Behavioral Assessment Scale Performance in Humans Influenced by Maternal Consumption of Environmentally Contaminated Lake Ontario Fish." *Journal of Great Lakes Research* 22, no. 2 (1996): 198–212.

Lopez-Martin, J. M., J. Ruiz-Olmo, and S. P. Minano. "Organochlorine Residue Levels in the European Mink (*Mustela Lutreola*) in Northern Spain." *Ambio* 3, nos. 4–5 (1994): 294–95.

"Louisiana Town Residents Exhibit Dangerous Levels of Dioxin." Greenpeace, no date. http://www.greenpeaceusa.org/features/mossville text.htm.

"Louisiana's Cancer Alley: An International Threat." Environment News Service, March 5, 1999. http://ens.lycos.com/ens/mar99/1999L-03-05-09.html.

Mably, Thomas A., et al. "*In Utero* and Lactational Exposure of Male Rats to 2,3,7,8-Tetrachlorodibenzo-P-dioxin. 1. Effects on Androgenic Status." *Toxicology and Applied Pharmacology* 114 (May 1992): 97–107.

Mably, Thomas A., D. L. Bjerke, R. W. Moore, A. Gendron-Fitzpatrick, and R. E. Peterson. "*In Utero* and Lactational Exposure of Male Rats to 2,3,7,8-tetrachlorodibenzo-p-dioxin. 2. Effects on Spermatogenesis and Reproductive Capability." *Toxicology and Applied Pharmacology* 114 (May 1992): 118–26.

Mably, Thomas A., R. W. Moore, D. L. Bjerke, and R. E. Peterson. "The Male Reproductive System Is Highly Sensitive to *in Utero* and Lactational TCDD Exposure." *Banbury Reports* 5 (1991): 69–78.

MacEachern, Frank, and Rachele Labrecque. "Clean-up Causes Health Fears." *Cornwall* (Ontario) *Standard-Freeholder*, July 13, 2001. http://www.standard-freeholder.southam.ca/.

Mansur, Michael. "After 15 Years, Dioxin Incineration at Times Beach, Mo. Is Finished." *Kansas City Star*, June 18, 1997. http://archive.nando times.com/newsroom/ntn/health/061897/health1_8068.html.

Marco, Gino J., Robert M. Hollingsworth, and William Durham, eds. *Silent Spring Revisited*. Washington, D.C.: American Chemical Society, 1987.

Mathews-Amos, Amy, and Ewann A. Berntson. "Turning Up the Heat: How Global Warming Threatens Life in the Sea." World Wildlife Fund and

Marine Conservation Biology Institute, 1999. http://www.world wildlife.org/news/pubs/wwf_ocean.htm.

McAndrew, Brian. "World Takes Aim at 'Dirty Dozen' Pollutants: Montreal Talks Bid to Ban Use of Worst Toxins." *Toronto Star,* June 29, 1998.

McClain, Mildred. "Food First Economics Bus Tour: More Testimonies." No date. http://www.foodfirst.org/bustour/testimonies2.html.

McDuffie, H. H., D. J. Klaassen, D. W. Cockcroft, and J. A. Dosman. "Framing and Exposure to Chemicals in Male Lung-Cancer Patients and Their Siblings." *Journal of Occupational Medicine* 30 (1988): 55–59.

McFarling, Usha Lee. "Fear Growing over a Sharp Climate Shift." *Los Angeles Times,* July 13, 2001.

———. "Scientists Warn of Losses in Ozone Layer over Arctic." *Los Angeles Times,* May 27, 2000.

McGinn, Anne Platt. "POPs Culture." *World Watch,* March–April 2000, 26–36.

McLachlan, J. A., ed. *Estrogens in the Environment.* New York: Elsevier North-Holland, 1980.

———. *Estrogens in the Environment II: Influences on Development.* New York: Elsevier North-Holland, 1985.

McLachlan, J. A., and K. S. Korach, eds. *Estrogens in the Environment III: Global Health Implications.* Washington, D.C.: National Institutes of Health and National Institute of Environmental Health Sciences, 1995.

Merchant, Carolyn. *The Death of Nature.* New York: Harper and Row, 1980.

Meunier, Bernard. "Catalytic Degradation of Chlorinated Phenols." *Science* 296 (April 12, 2002): 270–71.

Mills, P. K., G. R. Newell, and D. E. Johnson. "Testicular Cancer Associated with Employment in Agricultural and Oil and Natural-gas Extraction." *Lancet* (1984): 207–10.

Mocarelli, Paolo, Pier Mario Gerthoux, Enrica Ferrari, Donald G. Patterson Jr., Stephanie M. Kieszak, Paolo Brambilla, Nicoletta Vincoli, Stefano Signorini, Pierluigi Tramacere, Vittorio Carreri, Eric J. Sampson, Wayman E. Turner, and Larry L. Needham. "Paternal Concentrations of Dioxin and Sex Ratio of Offspring." *Lancet* 355 (May 27, 2000): 1858–63. http://irptc.unep.ch/pops/newlayout/press_items.htm.

Mofina, Rick. "Study Pinpoints Dioxin Origins: Cancer-Causing Agents in Arctic Aboriginals' Breast Milk Comes from U.S. and Quebec." *Montreal Gazette,* October 4, 2000.

Montague, Peter. "Fish Sex Hormones." *Rachel's Environment and Health News* 545, May 8, 1997. http://www.rachel.org/bulletin/index.cfm? St=2.

———. "Frogs, Alligators, and Pesticides." *Rachel's Environment and Health News* 590, March 19, 1998. http://www.rachel.org/bulletin/index. cfm?St=2.

————. "How We Got Here, Part 1; The History of Chlorinated Diphenyl (PCBs)." *Rachel's Environment and Health News* 327, March 4, 1993. Available at http://www.rachel.org/bulletin/bulletin.cfm?Issue_ID= 802&bulletin_ID=48.

————. "A New Era in Environmental Toxicology." *Rachel's Environment and Health News* 365, November 25, 1993. http://www.rachel.org/bulletin/index.cfm?St=2.

————. "PCBs Diminish Penis Size." *Rachel's Environment and Health News* 372, January 13, 1994. http://www.rachel.org/bulletin/index.cfm? St=2.

————. "The Weybridge Report." *Rachel's Environment and Health News* 547, May 22, 1997. http://www.rachel.org/bulletin/index.cfm?St=2.

Moriarty, F., ed. *Organochlorine Insecticides: Persistent Organic Pollutants.* London: Academic Press, 1975.

Moses, Alan. "Quality of Sperm Unchanged over 50 Years." Reuters News Service, February 29, 2000. Accessed at Web site of the Chlorine Chemistry Council. http://c3.org/news_center/third_party/02-29-00. html.

Moses, Marion. "Pesticides and Breast Cancer." *Pesticides News* 22 (December 1993): 3–5.

Moyers, Bill. "Trade Secrets: A Moyers Report" Program transcript. Public Broadcasting Service. March 26, 2001. http://www.pbs.org/trade secrets/transcript.html.

Murphy, Kim. "Front-Row Exposure to Global Warming: Engineers Say Alaskan Village Could Be Lost As Sea Encroaches." *Los Angeles Times,* July 8, 2001.

Nance, John J. *What Goes Up: the Global Assault on Our Atmosphere.* New York: William Morrow, 1991.

Napier, Robert. "Hot Air on the Environment." *Guardian* (London), August 16, 2001.

National Academy of Sciences National Research Council. *Hormonally Active Agents in the Environment.* Washington, D.C.: National Academies Press, 1999.

National Toxicology Program. *Endocrine Disruptors Low-Dose Peer Review Report.* Research Triangle Park, N.C.: National Institute of Environmental Health Sciences. October 10–12, 2000. http://ntp-server. niehs.nih.gov/htdocs/liason/lowdosewebpage.html.

Negoita, S., L. Swamp, B. Kelley, and D. O. Carpenter. "Chronic Diseases Surveillance of St. Regis Mohawk Health Service Patients." *Journal of Public Health Management Practice* 7, no. 1 (2001): 84–91.

Nelson, Bryn. "Frogs Feel Effect of a Herbicide; Sexual Damage Includes Loss of Voice in Males." *Newsday,* April 16, 2002.

Netting, Jessa. "Pesticides Implicated in Declining Frog Numbers." *Nature* 408 (December 4, 2000): 760.

Neubert, R., G. Golor, R. Stahlman, H. Helge, and D. Neubert. "Polyhalogenated Dibenzo-p-dioxins and Dibenzo-furans and the Immune System. 4. Effects of Multiple-Dose Treatment with 2,3,7,8-tetrachlorodibenzo-p-dioxin (TCDD) on Peripheral Lymphocyte Subpopulations of a Non-human Primate (*Calloithrix Jacchus*)." *Archives of Toxicology* 66 (1992): 250–71.

Newman, Paul A. "Preserving Earth's Stratosphere." *Mechanical Engineering*, October 1998. http://www.memagazine.org/backissues/october 98/features/stratos/stratos.html.

Norstrom, R. J., and D. C. G. Muir. "Chlorinated Hydrocarbon Contaminants in Arctic Marine Mammals." *Science of the Total Environment* 154 (1994): 107–28.

North American Commission for Environmental Cooperation. "Study Links Dioxin Pollution in Arctic to North American Sources." October 4, 2000. http://cec.org.

Nuttall, Nick. "Global Warming Boosts el Niño." *Times* (London), October 26, 2000.

O'Brien, R. D., and Izuru Yamamoto, eds. *Biochemical Toxicology of Insecticides.* New York: Academic Press, 1970.

O'Neill, Annie. "Damaged Lives: Vietnamese Veterans and Children: While World Leaders Debate the Effects of Agent Orange, a Multinational Project Reaches Out to People at the Center of the Storm." *Pittsburgh Post-Gazette,* November 5, 2000. http://groups.yahoo.com/group/ VeteranIssues/message/364.

Ouellet, Martin, et al. "Hind-limb Deformities (*Ectromelia, Ectrodactyly*) in Free-Living Anurans from Agricultural Habitats," *Journal of Wildlife Diseases* 33 (1997): 95–104.

Overpeck, J. T. "Warm Climate Surprises." *Science* 271 (March 29, 1996): 1820.

Palter, Jay, and Jack Weinberg. *Just a Matter of Time: Corporate Pollution and the Great Lakes.* Washington, D.C.: Greenpeace International Chlorine-Free Campaign, July 1994. http://www.greenpeace.org/ ~toxics/reports/reports.html.

Paulozzi, L. J., J. D. Erickson, and R. J. Jackson. "Hypospadias Trends in Two U.S. Surveillance Systems." *Pediatrics* 100 (1997): 831–34.

"PD 2000 Projects Arctic Monitoring and Research—Project Directory." April 11, 2001. http://amap.no/pd2000.htm.

*Persistent Organic Pollutants: Considerations for Global Action: IFCS Experts Meeting on POPs Final Report.* Intergovernmental Forum on Chemical Safety, Manila, Philippines, June 1996.

"Persistent Organic Pollutants in Asia: An Ongoing Disaster." Greenpeace International, November 10, 1998. http://www.greenpeace.org/press releases/toxics/1998nov10.html.

Pesticides Trust. "Persistent Organic Pollutants and Reproductive Health." From a briefing for UNISON prepared by the Pesticides Trust, London, 1999. http://irptc.unep.ch/pops/default.html.

Peterson, R. E., H. M. Theobald, and G. L. Kimmel. "Developmental and Reproductive Toxicity of Dioxins and Related Compounds: Cross-Species Comparisons." *Critical Reviews in Toxicology* 23, no. 3 (1993): 283–355.

Pianin, Eric. "Toxic Chemical Review Process Faulted; Scientists on EPA Advisory Panels Often Have Conflicts of Interest, G.A.O. Says." *Washington Post*, July 16, 2001. http://www.washingtonpost.com/wp-dyn/articles/A59494-2001Jul13.html.

Pierson, Nova. "Toxic Travel Fears." *Calgary Sun*, January 17, 2000.

Pimentel, David, and Hugh Lehman, eds. *The Pesticide Question: Environment, Economics, and Ethics.* New York: Chapman and Hall, 1993.

Polakovic, Gary. "Earth Losing Air-Cleansing Ability, Study Says; Worldwide Decline in a Molecule That Fights Pollution Is Found, but Experts Call the Losses Slight and Not Alarming." *Los Angeles Times*, May 4, 2001.

Poland, A., D. Palen, and E. Glover. "Tumor Promotion by TCDD in Skin of HRS/3 Hairless Mice." *Nature* 300 (1982): 271–73.

"POPs Invade Far Reaches of the Earth." Environment News Service. August 12, 1999. http://ens.lycos.com/ens/aug99/1999L-08–12–04.html.

"POPs Treaty Goes to U.S. Senate for Ratification." Environment News Service, April 11, 2002. http://ens-news.com/ens/apr2002/2002L-04-11-02.html.

Posten, Lee. "National Academy of Sciences Report on Hormone Disruptors Released; Growing Evidence of Damaging Health Effects Renews Call for Increased Research into Toxic Chemical Threat." Washington, D.C.: World Wildlife Fund, August 4, 1999. http://www.world wildlife.org/toxics/whatsnew/pr_7.htm.

Powell, Michael. "EPA Orders Record PCB Cleanup; GE to Foot $480 Million Bill to Dredge Upper Hudson River." *Washington Post*, August 2, 2001.

Pugliese, David. "An Expensive Farewell to Arms: The U.S. Has Abandoned 51 Military Sites in Canada; Many Are Polluted, and Taxpayers Are Paying Most of the $720 Million Cleanup Cost." *Montreal Gazette*, April 28, 2001.

Purdom, C., et al. "Estrogenic Effects of Effluents from Sewage Treatment Works." *Chemistry and Ecology* 8 (1994): 275–85.

Raj, Ranjit Dev. "Vanishing Vultures Bode Ill for Indians." Interpress Service, January 3, 2000. http://irptc.unep.ch/pops/newlayout/press_items.htm.

Rawn, Dorothea F. K., Derek C. G. Muir, Dan A. Savoie, G. Bruno Rosenberg, W. Lyle Lockhart, and Paul Wilkinson. "Historical Deposition of

PCBs and Organochlorine Pesticides to Lake Winnipeg (Canada)." *Journal of Great Lakes Research* 26 (2000): 3–17.

Reece, E. R. "Ontogeny of Immunity in Inuit Infants." *Arctic Medical Research* 45 (1987): 62–66.

Reggiani, G. "Acute Human Exposure to TCDD [Dioxins] in Seveso, Italy." *Journal of Toxicology and Environmental Health* 6 (1980): 27–43.

Reinjders, P. "Reproductive Failure in Common Seals Feeding off Fish from Polluted Coastal Waters." *Nature* 324 (1986): 456–57.

*Report on Carcinogenesis: TCDD.* Bethesda, Md.: National Institute of Environmental Health, National Toxicology Program, 1998.

"A Report by the World Wildlife Fund, the Wisconsin Potato and Vegetable Growers Association, and the Integrated Pest Management Team, at the University of Wisconsin." July 18, 2000. http://www. worldwildlife.org/toxics/whatsnew/pr_17.htm.

Revkin, Andrew C. "Study Finds a Decline in Natural Air Cleanser." *New York Times,* May 4, 2001.

———. "EPA Sharply Curtails the Use of a Common Insecticide." *New York Times,* June 9, 2000. http://irptc.unep.ch/pops/newlayout/press_ items.htm.

Rice, D. C., and S. Hayward. "Effects of Postnatal Exposure to a PCB Mixture in Monkeys on Non-spatial Discrimination Reversal and Delayed Alternation Performance." *Neurotoxicology* 18, no. 2 (1997): 479–94.

Rier, S. E., D. C. Martin, R. E. Bowman, W. P. Dmowski, and J. L. Becker. "Endometriosis in Rhesus Monkeys (*Macaca Mulatta*) Following Chronic Exposure to 2,3,7,8-tetrachlordibenzo-p-dioxin." *Fundamentals of Applied Toxicology* 21 (1993): 433–41.

Ries, Lynn A. G., Phyllis A. Wingo, Daniel S. Miller, Holly L. Howe, Hannah K. Weir, Harry M. Rosenberg, Sally W. Vernon, Kathleen Cronin, and Brenda K. Edwards. "The Annual Report to the Nation on the Status of Cancer, 1973–1997, with a Special Section on Colorectal Cancer." *Cancer* 88, no. 10 (2000): 2398–424. http://seer.cancer.gov/publica tions/csr1973_1998/overview/.

Risebrough, R. W., and B. de Lappe. "Accumulation of Polychlorinated Biphenyls in Ecosystems." *Environmental Health Perspectives* 1 (1972): 39–45.

Ritter, L., K. R. Solomon, J. Forget, M. Stemeroff, and C. O'Leary. *A Review of Selected Persistent Organic Pollutants.* New York: United Nations International Program on Chemical Safety, December 1995.

Robertson, Larry W., and Larry G. Hansen, eds. *PCBs: Recent Advances in Environmental Toxicology and Health Effects.* Lexington: University Press of Kentucky, 2001.

Robock, Alan. "Pinatubo Eruption: The Climatic Aftermath." *Science* 295 (February 15, 2002): 1242–44.

Rogan, Walter J., et al., "Congenital Poisoning by Polychlorinated Biphenyls and Their Contaminants in Taiwan." *Science* 241 (July 15, 1988): 334–36.

Roslin, Alex. "Crees Revive Hydro Project." *Montreal Gazette,* January 21, 2000.   http://www.montrealgazette.com/news/pages/010121/50367 05.html.

Rowland, Sherwood, and Mario Molina. "Stratospheric Sink for Chlorofluoromethanes: Chlorine Atom-Catalyzed Destruction of Ozone." *Nature* 249 (June 28, 1974): 810–12.

Rudd, Robert L. *Pesticides and the Living Landscape.* Madison: University of Wisconsin Press, 1964.

Rugman, F. P., and R. Cosstick. "Aplastic Anaemia Associated with Organochlorine Pesticides: Case Reports and Review of Evidence." *Journal of Clinical Pathology* 43 (1990): 98–101.

Rupa, D. S., P. P. Reddy, and O. S. Reddy. "Reproductive Performance in Population Exposed to Pesticides in Cotton Fields in India." *Environmental Research* 55 (1991): 123–28.

"Rural Men Found to Have Poorer Semen Quality." Associated Press in *Omaha World-Herald,* November 11, 2002.

Rusnell, Charles. "U.S. Military Wastes Entering Canada; Ottawa Concerned with Political Fallout, Document Shows." *Edmonton Journal,* March 31, 2000.

Russell, Edmund P. III. "Speaking of Annihilation: Mobilizing for War against Human and Insect Enemies, 1914–1945." *Journal of American History* 82, no. 4 (1996): 1505–29.

Russell, Ronald W., et al. "Organochlorine Pesticide Residues in Southern Ontario Spring Peepers." *Environmental Toxicology and Chemistry* 14 (1995): 815–17.

Russell, Ronald W., and Stephen J. Hecnar. "The Ghosts of Pesticides Past?" *Froglog* 19 (November 1996): 1.

Russell, Ronald W., et al. "Polychlorinated Biphenyls and Chlorinated Pesticides in Southern Ontario, Canada, Green Frogs." *Environmental Toxicology and Chemistry* 16 (1997): 2258–63.

Russell-Jones, Robin. "Letter: Ozone in Peril." *London Independent,* December 7, 2000.

Rylander, Lars, Ulf Stromberg, and Lars Hagmar. "Decreased Birthweight in Infants Born to Women with a High Dietary Intake of Fish Contaminated with Persistent Organochlorine Compounds." *Scandinavian Journal of Work and Environmental Health* 21 (1995): 368–75.

———. "Dietary Intake of Fish Contaminated with Persistent Organochlorine Compounds in Relation to Birth-Weight." *Scandinavian Journal of Work and Environmental Health* 22 (1996): 260–66.

Saar, Robert A. "In the Backyard, A Potent Source of Pollution." *New York Times,* January 4, 2000. http://irptc.unep.ch/pops/newlayout/press_ items.htm.

Safe, S. H. "Environmental and Dietary Oestrogens and Human Health: Is There a Problem? *Environmental Health Perspectives* 103, no. 4 (1995): 346–51.

———. "Polychlorinated Biphenyls (PCBs): Environmental Impact, Biochemical and Toxic Responses, and Implications for Risk Assessment." *Critical Reviews in Toxicology* 24, no. 2 (1994): 87–149.

Santillo, David. "World Chemical Supplies Contaminated with Toxic Chemicals." Greenpeace Listserve, March 19, 2000. http://www. greenpeace.org/~toxics/reports/recipe.html.

Schafer, Kristin. "Nowhere to Hide: Persistent Toxic Chemicals in the U.S. Food Supply." Press release, Pesticide Action Network North America, San Francisco, California. November 2000. http://www.panna.org/ panna/resources/resources.html.

Schantz, S. L., A. M. Sweeney, J. C. Gardiner, H. E. B. Humphrey, R. J. McCaffrey, D. M. Gasior, K. R. Srikanth, and M. L. Budd. "Neuropsychological Assessment of an Aging Population of Great Lakes Fisheaters." *Toxicology and Industrial Health* 12 (1996): 403–17.

Schecter A., ed. *Dioxins and Health.* New York: Plenum, 1994.

Schecter, A., O. Papke, M. Ball, D. C. Hoang, C. D. Le, Q. M. Nguyen, T. Q. Hoang, N. P. Nguyen, K. Pham, C. Huynh, D. Vo, J. D. Constable, and J. Spencer. "Dioxin and Dibenzofuran Levels in Blood and Adipose Tissue of Vietnamese from Various Locations in Vietnam in Proximity to Agent Orange Spraying." *Chemosphere* 25, nos. 7–10 (1992): 1123–28.

Schecter, A., P. Toniolo, L. C. Dai, T. B. Thuy, and M. S. Wolffe. "Blood Levels of DDT and Breast-Cancer Risk among Women Living in the North of Vietnam." *Archives of Environmental Contamination and Toxicology* 33 (1997): 453–456.

Schecter, Arnold, Le Cao Dai, Olaf Papke, Joelle Prange, John D. Constable, Muneaki Matsuda, Vu Duc Thao, and Amanda L. Piskac. "Recent Dioxin Contamination from Agent Orange in Residents of a Southern Vietnamese City." *Journal of Occupational and Environmental Medicine* 43, no. 5 (May 2001): 435–43.

Schettler, Ted. "The Precautionary Principle and Persistent Organic Pollutants: The Science and Environmental Health Network." Press release, March 2000. http://www.alphacdc.com/ien/pops_precautionary_ted. html.

———. Statement at Willard Hotel, Washington, D.C., September 7, 2000. http://www.psr.org/trited.html.

Schettler, Ted, Gina Solomon, Maria Valenti, and Anne Huddle. *Generations at Risk: Reproductive Health and the Environment.* Cambridge, Mass.: MIT Press, 1999.

Schrope, Mark. "Successes in Fight to Save Ozone Layer Could Close Holes by 2050." *Nature* 408 (December 7, 2000): 627.

Schwanke, Jane, and Pamela Yoder. "Malibu Barbie, Holiday Barbie . . . Toxic Barbie? Some Vintage Toys May Ooze Chemical That Could Harm Kids." *WebMD Medical News*, August 25, 2000. http://content.health.msn.com/content/article/1728.60731.

Schwarcz, Joe. "Our Daily Dioxin: It's a Potent Carcinogen and You'll Probably Have Some for Supper." *Montreal Gazette Magazine*, January 14, 2001.

Schwetz, B. A., B. M. J. Leontg, and B. J. Gehring. (1975). "The Effect of Maternally Inhaled Trichloroethylene, Perchloroethylene, Methyl Chloroform, and Methyl Chloride on Embryonal and Fetal Development in Mice and Rats." *Toxicology and Applied Pharmacology* 32 (1975): 84–96.

"Scientists Report Large Ozone Loss." *USA Today*, April 6, 2000.

Seely, Hart. "Toxins Remain 18 Years Later: Landfill Near Massena Polluting Water Where Mohawk Children Played." *Syracuse Post-Standard*. June 24, 2001.

Sen Gupta, Sayam, Matthew Stadler, Christopher A. Noser, Anindya Ghosh, Bradley Steinhoff, Dieter Lenior, Colin P. Horwitz, Karl-Werner Schramm, and Terrence J. Collins. "Rapid Total Destruction of Chlorophenols by Activated Hydrogen Peroxide." *Science* 296 (April 12, 2002): 326–28.

Senanayake, Ranil, et al. "Frog Tea?" *Froglog* 23 (August 1997): 2.

Sengupta, Smini. "A Sick Tribe and a Dump As a Neighbor." *New York Times*, April 7, 2001. http://www.nytimes.com/2001/04/07/nyregion/07MOHA.html.

"Settlement Will Continue PCB Cleanup in Anniston." Environment News Service, March 25, 2002. http://ens-news.com/ens/mar2002/2002L-03-25-09.html.

Shane, B. S. "Human Reproductive Hazards." *Environmental Science and Technology* 23, no. 10 (1989): 1187–95.

Sharpe R. M. "Declining Sperm Counts in Men: Is There an Endocrine Cause?" *Journal of Endocrinology* 136 (1993): 357–60.

Sharpe, R. M., and N. E. Skakkebaek. "Are Oestrogens Involved in Falling Sperm Counts and Disorders of the Male Reproductive Tract?" *Lancet* 341 (1993): 1392–95.

"Shell Will Pay $28 Million to Clean Wells of MTBE." Environment News Service, August 6, 2002. http://ens-news.com/ens/aug2002/2002-08-06-09.asp.

Sherwood, Steven. "A Microphysical Connection among Biomass Burning, Cumulus Clouds, and Stratospheric Moisture." *Science* 295 (February 15, 2002): 1272–75.

Shindell, Drew T., David Rind, and Patrick Lonergan. "Increased Polar Stratospheric Ozone Losses and Delayed Eventual Recovery Owing to Increasing Greenhouse-Gas Concentrations." *Nature* 392 (April 9, 1998): 589–92.

"Shintech: The Battle Continues." *E: The Environmental Magazine,* March–April 1999. http://www.emagazine.com/march-april_1999/0399up dates.html.

Sierra Club. "Stories from the Field: Corporate Pollution: Shintech and Louisiana; Eyes of the World on 'Cancer Alley.'" No date. http://www.sierraclub.org/toxics/resources/shintech.asp.

Simmons, Ann M. "Tanzania Begins to Deal with Toxic Wastelands, Pesticides; With No Sound Disposal Method, Stockpiles Keep Growing. Experts Warn That Unless Quick Action Is Taken, the Situation Could Be Catastrophic." *Los Angeles Times,* March 30, 2000. http://irptc.unep.ch/pops/newlayout/press_items.htm.

Sinks, T., G. Steele, and A. B. Smith. "Mortality among Workers Exposed to Polychlorinated Biphenyls." *American Journal of Epidemiology* 136, no. 4 (1992): 389–98.

Siskind, David E., Stephen D. Eckermann, and Michael E. Summers, eds. *Atmospheric Science across the Stratopause.* Washington, D.C.: American Geophysical Union, 2000.

Skakkebaek, N. E. "Possible Carcinoma-*in-situ* of the Testis." *Lancet* 1 (7756) (September 1972): 516–57.

Skogland, T. "Radiocesium Concentrations in Wild Reindeer at Dovrefjell, Norway." *Rangifer* 7 (1987): 42–45.

Skrzycki, Cindy, and Joby Warrick. "EPA Links Dioxin to Cancer: Risk Estimate Raised Tenfold." *Washington Post,* May 17, 2000. http://irptc.unep.ch/pops/newlayout/press_items.htm.

Smoger, G. H., P. C. Kahn, G. C. Rodgers, and S. Suffin. "*In-utero* and Postnatal Exposure to 2,3,7,8-TCDD in Times Beach, Missouri." *Organohalogen Compounds* 13 (1993): 345–48.

"Smuggled CFCs Convict Florida Man." Environment News Service. January 11, 2002. http://ens-news.com/ens/jan2002/2002L-01-11-09.html.

Snow, Mitch, and John Zogorski. "Gasoline Additive Found in Urban Ground Water." United States Geological Survey, March 31, 1995. http://sd.water.usgs.gov/nawqa/vocns/mtbe.htm.

Soto, A. M., K. L. Chung, and C. Sonnenschein. "The Pesticides Endosulfan, Toxaphene, and Dieldrin Have Oestrogenic Effects on Human Estrogen-Sensitive Cells." *Environmental Health Perspectives* 102 (1994): 380–83.

Souder, William. "A Pesticide-Parasite Role in Frogs' Deformities?" *Washington Post,* July 15, 2002.

Spearow, J. L., P. Doemeny, et al. "Genetic Variation in Susceptibility to Endocrine Disruption by Estrogen in Mice." *Science* 285 (1999): 1259–61.

Squillace, Paul J., Daryll A. Pope, and Curtis V. Price. "Occurrence of Gasoline Additive MTBE in Shallow Ground Water in Urban and Agricultural Areas." U.S. Geological Survey Fact Sheet 114–95, 1995. http://wwwrvares.er.usgs.gov/nawqa/nawqa_home.html.

Stalling, D. L., L. M. Smith, J. D. Petty, J. W. Hogan, J. L. Johnson, C. Rappe, and H.-R. Buser. "Residues of Polychlorinated Dibenzo-p-dioxins and Dibenzofurans in Laurentian Great Lakes Fish." In R. E. Tucker, A. L. Young, and A. P. Gray. *Human and Environmental Risks of Chlorinated Dioxin and Related Compounds.* New York, Plenum, 1983.

"State, G.M. Talking So Lawsuit Is Set Aside." June 11, 2001. http://syracuse.com/newsflash/index.ssf?/cgi-free/getstory_ssf.cgi?n0505_BC_NY—contamination&news&newsflash-newyork-syr.

"State, Mohawks Threaten Lawsuit Unless G.M. Cleans Up," March 22, 2001. http://www.topica.com/lists/SSVOP.

Stein, Matthew. *When Technology Fails: A Manual for Self-Reliance and Planetary Survival.* Santa Fe, N.M.: Clear Light, 2000.

Steinhaus, Edward A. "Concerning the Harmlessness of Insect Pathogens and the Standardization of Microbial Control Products." *Journal of the Economics of Entomology* 50, no. 6 (December 1957): 715–20.

Sterling, P. *Sea and Earth: The Life of Rachel Carson.* New York: Dell, 1970.

Stevens, J. T., C. B. Breckenridge, L. T. Wetzel, J. H. Gillis, L. G. Luempert, and J. C. Eldridge. "Hypothesis for Mammary Tumorigenesis in Sprague-Dawley Rats Exposed to Certain Triazine Herbicides." *Journal of Toxicology and Environmental Health* 43 (1994): 139–53.

Stevens, William K. *The Change in the Weather: People, Weather, and the Science of Climate.* New York: Delacorte Press, 1999.

Stewart, Irvine, et al. "Evidence of Deteriorating Semen Quality in the United Kingdom: Birth Cohort Study in 577 Men in Scotland over 11 Years." *British Medical Journal* 312 (February 1996): 467–71.

Sumpter, John P. "Feminized Responses in Fish to Environmental Estrogens." *Toxicology Letters* 82–83 (December 1995): 737–42.

Sumpter, John P., and Susan Jobling. "Vitellogenesis As a Biomarker for Estrogenic Contamination of the Aquatic Environment." *Environmental Health Perspectives* 103, suppl. 7 (October 1995): 173–77.

Suskind, R. R., and V. S. Hertzberg. "Human Health Effects of 2,4,5-T and Its Toxic Contaminants." *Journal of the American Medical Association* 251 (1984): 2372–80.

Svensson, B., T. Hallberg, A. Nilsson, B. Akesson, A. Schutz, and L. Hagmar. "Immunological Competence and Liver Function in Subjects Consuming Fish with Organochlorine Contaminants." In *DIOXIN '93: 13th International Symposium on Chlorinated Dioxins and Related Compounds. Organohalogen Compounds,* vol. 13, edited by Frank H.

Fiedler, O. Hutzinger, W. Parzefall, A. Riss, and S. Safe. Vienna, Austria: Federal Environmental Agency, 1993.

Swan, Shanna H., Charlene Brazil, Erma Z. Drobnis, Fan Liu, Robin L. Kruse, Maureen Hatch, J. Bruce Redmon, Christina Wang, James W. Overstreet, and the Study for Future Families Research Group. "Geographic Differences in Semen Quality of Fertile U.S. Males." Environmental Health Perspectives Online, November 11, 2002. http://www. ehponline.org/swan2002.

Swan S. H., E. P. Elkin, and L. Fenster. "Have Sperm Densities Declined? A Re-analysis of Global Trend Data." *Environmental Health Perspectives* 105: 11 (1997): 1228–32.

Tabazadeh, A., K. Drdla, M. R. Schoeberl, P. Hamill, and O. B. Toon. "Arctic 'Ozone Hole' in a Cold Volcanic Stratosphere." *Proceedings of the National Academy of Sciences* 99, no. 5 (March 5, 2002): 2609–12.

Tabazadeh, A., E. J. Jensen, O. B. Toon, K. Drdla, and M. R. Schoeberl. "Role of the Stratospheric Polar Freezing Belt in Denitrification." *Science* 292 (March 30, 2001): 2591–94.

Tanabe, Shinsuke. "PCB Problems in the Future: Foresight from Current Knowledge." *Environmental Pollution* 50 (1988): 5–28.

Taskinen H., M-L. Lindbohm, and K. Hemminki. "Spontaneous Abortion among Women Working in the Pharmaceutical Industry." *British Journal of Industrial Medicine* 43 (1986): 199–205.

Taylor, Karin. "Alpine Lakes Trap 'Dirty Dozen' Poisons." Reuters News Service, April 10, 2000. http://irptc.unep.ch/pops/newlayout/press_items.htm.

*The Third Citizens' Conference on Dioxin and Other Synthetic Hormone Disrupters, March 15–17, 1996, Baton Rouge, Louisiana.* Sanford, N.C.: Citizens' Conference on Dioxin, Inc. http://www.workonwaste.org/wastenots/wn357.htm.

Thomas, Fred. "Clear-Cut Answers on Safety of Omaha's Drinking Water in Short Supply." *Omaha World-Herald*, August 27, 1995.

Thomas, Katie. "Toxic Threats to Tribal Lands." *Newsday,* March 25, 2001. http://www.newsday.com/coverage/current/news/sunday/nd8399.htm.

Thomas, Kristin Bryan, and Theo Colborn. "Organo-chlorine Endocrine Disruptors in Human Tissue." *Advances in Modern Environmental Toxicology* 21 (1992): 342–43.

Thomas, P. T. "Approaches Used to Assess Chemically Induced Impairment of Host Resistance and Immune Function." *Toxic Substances Journal* 10 (1990): 241–78.

Thompson, Elizabeth. "A Slimmer You May Be Less Healthy: Quebec Researchers Find Link between Weight Loss and Higher Levels of Pollutants in the Body." *Montreal Gazette*, January 29, 2001.

Thornton, Joe. *Pandora's Poison: Chlorine, Health, and a New Environmental Strategy.* Cambridge, Mass.: MIT Press, 2000.

———. "PVC: The Poison Plastic." Greenpeace USA, Washington, D.C., April 1997. http://www.greenpeace.org/~usa/reports/toxics/PVC/cradle/dc gsum.html.

Thrupp, L. A. "Sterilization of Workers from Pesticide Exposure: The Causes and Consequences of DBCP-Induced Damage in Costa Rica and Beyond." *International Journal of Health Services* 21, no. 4 (1991): 731–57.

Tilson, Hugh A., et al. "Polychlorinated Biphenyls and the Developing Nervous System: Cross-Species Comparisons." *Neurotoxicology and Teratology* 12 (1990): 239–48.

Toivonen, T., et al. "Parathion Poisoning Increasing Frequency in Finland." *Lancet* (1959): 175–76.

Tolbert, Margaret A., and Owen B. Toon. "Solving the P[olar] S[tratospheric] C[loud] Mystery." *Science* 292 (April 6, 2001): 61–63.

Toniolo, P. G., M. Levitz, and A. Zeleniuch-Jacquotte. "A Prospective Study of Endogenous Estrogens and Breast Cancer in Post-menopausal Women." *Journal of the National Cancer Institute* 87 (1995): 190–97.

"Toxic Waste in Japan: The Burning Issue." *Economist* (London), July 25, 1998.

"Toxics Found in Snowcaps." Associated Press, October 5, 1999. http://www.science.uottawa.ca/~biologie/PROFS/jblais.htm#publications.

Trichopoulos, D. "Hypothesis: Does Breast Cancer Originate *in Utero*?" *Lancet* 335 (1990): 939–40.

"$2.2 Million Supports Phytoremediation Studies." Environment News Service, March 12, 2002 http://ens-news.com/ens/mar2002/2002L-03-12-09.html.

Tyler, Michael J. "Herbicides Kill Frogs." *Froglog* 21 (March 1997): 2.

"U.N. Agency Calls for Faster Disposal of Toxic Pesticide Waste Stocks." May 9, 2001. www.fao.org.

U.N. Environment Program. *International Action to Protect Human Health and the Environment through Measures Which Will Reduce and/or Eliminate Emissions and Discharges of Persistent Organic Pollutants, Including the Development of an International Legally Binding Instrument.* Governing Council Decision 19/13 C, February 1997.

U.N. Environmental Program. *Persistent Organic Pollutants.* Governing Council Decision 18/32, May 1995.

U.N. Environmental Program. *Report of the Intergovernmental Conference to Adopt a Global Program of Action for the Protection of the Marine Environment from Land-Based Activities.* United Nations Environment Program, December 1995.

U.S. Environmental Protection Agency. *EPA Draft Dioxin Reassessment.* 2000. http://www.epa.ncea/dioxin.htm.

U.S. Environmental Protection Agency. "Fact Sheet: Persistent, Bioaccumulative and Toxics Initiative." August 1, 2002. http://www.epa.gov/
opptintr/pbt.

U.S. Environmental Protection Agency. *Federal Register* 63, no. 83 (April
30, 1998): 23785–86, cited in Cray and Harden, 1998.

U.S. Environmental Protection Agency. *Health Assessment Document for
2,3,7,8-tetrachlorodibenzo-p-dioxin (TCDD) and Related Compounds.*
3 Vols. Washington, D.C.: U.S. EPA, Office of Research and Development, 1994.

U.S. Environmental Protection Agency. *Public Health Implications of PCB
Exposures.* Atlanta: Agency for Toxic Substances and Disease. De-
cember 1996. http://www.epa.gov/region5/foxriver/lower_fox_river_
PCB_Exposures.htm.

U.S. Environmental Protection Agency. "Times Beach Settlement Reached."
Press release, July 20, 1990. http://www.epa.gov/history/topics/
times/01.htm.

"U.S. EPA Takes Action to Clean Great Lakes of Toxicity." Washington, D.C.,
November 6, 2000. http://www.planetark.org/dailynewsstory.cfm?
newsid=8799.

U.S. General Accounting Office. "Environmental Protection Agency's Science Advisory Board Panels: Improved Policies and Procedures
Needed to Ensure Independence and Balance." July 2001. http://
www.gao.gov/new.items/d01536.pdf.

U.S. Geological Survey. "MTBE in Ground Water." April 15, 1997. http://
sd.water.usgs.gov/nawqa/vocns/mtbe.htm.

"U.S. Government Turns Its Back on Dioxin Elimination in Global Pollution
Treaty." Greenpeace, 1999. http://www.greenlink.org/public/hot
issues/dioxin.html.

Urquhart, Frank. "Blue Whale Close to Extinction." *Scotsman,* July 19,
2001.

"UV Radiation Linked to Deformed Amphibians." Environment News Service, June 21, 2002. http://ens-news.com/ens/jun2002/2002-06-21-
09.asp#anchor4.

Van Emden, Helmut F., and David B. Peakall. *Beyond Silent Spring: Integrated Pest Management and Chemical Safety.* London: Chapman
and Hall and United Nations Educational Program, 1996.

Verrengia, Joseph B., and Tini Tran. "Vietnam's Children Feeling Effects of
Agent Orange." *Amarillo* [Texas] *Globe-News,* November 20, 2000.
http://www.amarillonet.com/stories/112000/hea_agentorange.shtml.

Versar, Inc. *Exposure Assessment for Incidentally-Produced Polychlorinated
Biphenyls (PCBs).* Draft Final Report, vol. 3. U.S. EPA Contract
No. 68-01-6271, August 15, 1983.

"Volcanic Eruptions Could Damage Ozone Layer." Environment News Service, March 5, 2002. http://ens-news.com/ens/mar2002/2002L-03-
05-09.html.

vom Saal, F. S., B. G. Timms, et al. "Prostate Enlargement in Mice Due to Fetal Exposure to Low Doses of Estradiol or Diethylstilbestrol and Opposite Effects at High Doses." *Proceedings of the National Academy of Sciences* 94 (1997): 2056–61.

Von Hernandez, Jayaraman N. *Toxic Legacies; Poisoned Futures: Persistent Organic Pollutants in Asia.* Washington, D.C.: Greenpeace International, November 1998.

Waddell, Craig, ed. *And No Birds Sing: Rhetorical Analysis of Rachel Carson's "Silent Spring."* Carbondale: Southern Illinois University Press of America, 2000.

Waller D. P., C. Presperin, M. L. Drum, A. Negrusz, A. K. Larsen, H. van der Ven, and J. Hibbard. "Great Lakes Fish As a Source of Maternal and Fetal Exposure to Chlorinated Hydrocarbons." *Toxicology and Industrial Health* 12 (1996): 335–45.

Watt-Cloutier, Sheila, and Terry Fenge. "Commentary: Impasse at POPs Talks Unacceptable for Inuit." *Nunatsiaq News,* March 17, 2000. http://www.nunatsiaq.com/archives/nunavut000331/nvt20317_24. html.

Webb, Jason. "Small Islands Say Global Warming Hurting Them Now." Reuters News Service, 1998. http://bonanza.lter.uaf.edu/~davev/nrm 304/glbxnews.htm.

Weisglas-Kuperus, N., T. C. J. Sas, C. Koopman-Esseboom, C. W. van der Zwan, M. A. J. de Ridder, A. Beishuizen, H. Hooijkaas, and P. J. J. Sauer. "Immunologic Effects of Background Prenatal and Postnatal Exposure to Dioxins and Polychlorinated Biphenyls in Dutch Infants." *Pediatric Research* 38, no. 3 (1995): 404–10.

Wetzel L. T., L. G. Luempert III, M. O. Breckenridge, J. T. Stevens, A. K. Thakur, P. J. Extrom, and J. C. Eldridge. "Chronic Effects of Atrazine on Estrus and Mammary Tumor Formation in Female Sprague-Dawley and Fischer 344 Rats." *Journal of Toxicology and Environmental Health* 43 (1994): 169–82.

"Whale and Dolphin Meat Sold in Japan Has High Levels of Dioxin." *Japanese Times Online,* July 4, 2000. http://www.japantimes.co.jp/cgi bin/getarticle.pl5?nn20000704a4.htm.

Whelan, Elizabeth M. "Who Says PCBs Cause Cancer?" *Wall Street Journal,* December 12, 2000.

Whitten, J. L. *That We May Live.* Princeton, N.J.: D. Van Nostrand, 1966.

Whorton, M. D., and D. E. Foliart. "Mutagenicity, Carcinogenicity, and Reproductive Effects of Dichloropropane (DBCP)." *Mutation Research* 123 (1983): 13–30.

Whorton, D., R. M. Krauss, S. Marshall, and T. M. Milby. "Infertility in Male Pesticide Workers." *Lancet* 2 (1259) (December 17, 1977): 1259–61.

Wigglesworth, V. D. "A Case of DDT Poisoning in Man." *British Medical Journal* 1 (April 14, 1945): 517–21.

Wilkinson, Clive, Olof Linden, Herman Cesar, Gregor Hodgson, Jason Rubens, and Alan E. Strong. "Ecological and Socioeconomic Impacts of 1998 Coral Mortality in the Indian Ocean: An ENSO Impact and a Warning of Future Change?" *Ambio* 28, no. 2 (March 1999): 188–96.

Williams, Wendy. "Pirate Fear." *Scientific American,* October 1999, 26.

Wipf, H. K., and J. Schmid. "Seveso: An Environmental Assessment." *Environmental Science Research* 26 (1983): 255–74.

Withgott, Jay. "Ubiquitous Herbicide Emasculates Frogs." *Science* 296 (April 19, 2002): 447–48.

"With Public's 'Right-To-Know' in Jeopardy, Greenpeace Kicks Off Bus Tour of Louisiana's Worst Chemical 'Hot Spots.'" Greenpeace, June 22, 1999. http://www.commondreams.org/pressreleases/june99/06 2299e.htm.

Wolfe W. H., J. E. Michalek, and J. C. Miner. "Determinants of TCDD Half-life in Veterans of Operation Ranch Hand." *Journal of Toxicology and Environmental Health* 41 (1994): 481–88.

Wolfe, W. H., J. E. Michalek, and J. C. Miner. "Paternal Serum Dioxin and Reproductive Outcomes among Veterans of Operation Ranch Hand." *Epidemiology* 6, no. 1 (1995): 17–22.

Wong K. C., and M. Y. Huang. "Children Born to PCB-Poisoned Mothers." *Clinical Medicine* 7 (1981): 83–87.

Woodhouse, C. A., and J. T. Overpeck. "2000 Years of Drought Variability in the Central United States." *Bulletin of the American Meteorological Society* 79, no. 12 (December 1998): 2693–714.

World Wildlife Fund. "Toxics—What's New." 2000. http://www.world wildlife.org/toxics/whatsnew/pr_7.htm.

Wright, Bruce S. "Woodcock Reproduction in DDT-Sprayed Areas of New Brunswick." *Journal of Wildlife Management* 24, no. 4 (1960): 419–20.

Wright, Lawrence. "A Reporter at Large: Silent Sperm." *New Yorker,* January 15, 1996, 42–47.

Yang, Y. G., H. Lebrec, and G. R. Burleson. "Effect of 2,3,7,8-tetrachlorodibenzo-p-dioxin (TCDD) on Pulmonary Influenza Virus Titer and Natural Killer Activity in Rats." *Toxicology and Applied Pharmacology* 23 (1994): 125–31.

Yao, Y., A. Hoffer, C. Chang, and A. Puga. "Dioxin Activates HIV-1 Gene Expression by an Oxidative Stress Pathway Requiring a Functional Cytochrome P450 CYP1A1 Enzyme." *Environmental Health Perspectives* 103, no. 4 (1995): 366–71.

Yassi, A., R. Tate, and D. Fish. "Cancer Mortality in Workers Employed at a Transformer Manufacturing Plant." *American Journal of Industrial Medicine* 25, no. 3 (1994): 425–37.

York, Geoffrey. "Russian City Ravaging Arctic Land." *Toronto Globe and Mail,* July 25, 2001.

Zieler, S., L. Feingold, R. A. Danley, and G. Craun. "Bladder Cancer in Massachusetts Related to Chlorinated and Contaminated Drinking Water: A Case-Control Study." *Archives of Environmental Health* 43, no. 2 (1988): 195–200.

Zook, D. R., and C. Rappe. "Environmental Sources, Distribution, and Fate of Polychlorinated Dibenzodioxins, Dibenzofurans, and Related Organochlorides." In *Dioxins and Health,* edited by A. Schecter. New York: Plenum, 1994.

# Index

Abortion, spontaneous, and PCBs, 180–81

Agarwal, Anil, 152

Agent Orange, 119, 237, 249; birth deformities and, 26; children and, 27, 29–30; sexual dysfunction in men, 208–9; use during Vietnam War, 24-31, 174

Akeya, Alexander, 58

Akwesasne, Mohawk reservation, 111–16; derivation of name, 111; diseases and PCBs at, 113–15; fluoride poisoning of cattle at, 115–16; origins of, 111; PCB contamination at, 111–16

Aldrin, 15, 249

Alkylphenols, 213–14; health damage, 214; uses, 214

Allen, Joe, 108

Allergies, in the Arctic, 65

Alligators and POPs, 149–51

Amato, Chris, 116

American Cyanamid Corp., 158

American Library Association, chlorine-free paper, 240–41

Animashaun, Kishi, 121

Anniston, Alabama, and PCBs, 235, 236

Apopka, Lake, alligators and POPs, 149–51

Arctic, as sink for POPs, 4, 5, 15, 17, 47–56, 65, 136, 252; global warming and, 57–68; hunting, and global warming, 60–61; mammals and POPs, generally, 136–37; mercury in, 68–69; stratospheric ozone, depletion, 69–71, 84, 85, 86, 88–89, 100; thunderstorms, 66; ultraviolet radiation, 70–71; volcanic aerosols and ozone loss, 100–101

Argentina, Newfoundland; U.S. Navy base, contamination of, 54

Armyworms, 157–58

Arsenic standards, drinking water, 237

Asbestos, clothing from, 1

Atrazine, 143–46, 186–87, 249; banned in European countries, 144; frogs and, 143–46; levels in Midwest U.S. water, 186

Austin, J., 83

Auyuittuq National Park (Baffin Island), 61

Avants, Liz, 122

Ayotte, Pierre, 181, 182

Bald eagles, POPs, 155; reproductive failure, 135

Barbie dolls, 221, 222

Basel Convention (1989), 233

Beckman, Kimberlee, 148

Belgium, chickens, dioxins, and PCBs (1999), 38–40

Beluga whales, PCBs in, 138–39; toxicity and diseases, 139

# About the Author

BRUCE E. JOHANSEN is Robert T. Reilly Professor of Communication and Native American Studies at the University of Nebraska at Omaha. His last book was *The Ecocide of Native America* (1995).